EDUCATION FOR A DIGITAL WORLD
Present Realities and Future Possibilities

EDUCATION FOR A DIGITAL WORLD

Present Realities and Future Possibilities

Edited By

Rocci Luppicini, PhD

Associate Professor, Department of Communication,
University of Ottawa, Canada; Editor in Chief,
International Journal of Technoethics

A. K. Haghi, PhD

Associate member of University of Ottawa, Canada;
Freelance Science Editor, Montréal, Canada

Apple Academic Press

TORONTO   NEW JERSEY

© 2013 by
Apple Academic Press Inc.
3333 Mistwell Crescent
Oakville, ON L6L 0A2
Canada

Apple Academic Press Inc.
1613 Beaver Dam Road, Suite # 104
Point Pleasant, NJ 08742
USA

First issued in paperback 2021

Exclusive worldwide distribution by CRC Press, a Taylor & Francis Group

ISBN 13: 978-1-77463-198-0 (pbk)
ISBN 13: 978-1-926895-10-9 (hbk)

Library of Congress Control Number: 2012935660

Library and Archives Canada Cataloguing in Publication

Education for a digital world: present realities and future possibilities/edited by Rocci Luppicini, A.K. Haghi.

Includes bibliographical references and index.
ISBN 978-1-926895-10-9
1. Educational technology. I. Luppicini, Rocci, 1970- II. Haghi, A. K

LB1028.3.E25 2012   371.33   C2011-908706-5

Apple Academic Press also publishes its books in a variety of electronic formats. Some content that appears in print may not be available in electronic format. For information about Apple Academic Press products, visit our website at **www.appleacademicpress.com**

Contents

List of Contributors

Jessica Backen
University of Ottawa, Canada

Patricia Barber
University of Ottawa, Canada

A. K. Haghi
University of Guilan, Iran

Anton van Hamel
University of Ottawa, Canada

Xue Lin
University of Ottawa, Canada

Rocci Luppicini
University of Ottawa, Canada

M. Mahdavi
University of Guilan, Iran

Janine Moir
University of Ottawa, Canada

Carol Myhill
University of Ottawa, Canada

B. Noroozi
University of Guilan, Iran

List of Abbreviations

ASP	Active server pages
ACS	American chemical society
AU	Assumption University
CIPA	Children's Internet Protection Act
CVEs	Collaborative virtual environments
CFD	Computational Fluid Dynamics
CBT	Computer-based training
CSCL	Computer-supported collaborative learning
DOPA	Deleting Online Predators Act
E-learning	Electronic learning
ECPD	Engineers Council for Professional Development
EMIT	European Medical Imaging Technology
GUI	Graphical user interface
HTML	Hyper Text Markup Language
ICT	Information and communication technologies
ISS	Information systems success
ID	Instructional Design
LAMS	Learning Activity Management Systems
LAN	Local area networks
LMSs	Learning management systems
MEP	Medical Physics and Engineering
MIL	Multiple instances learning
MUDs	Multi-User Domains
NI	National Instruments
PHP	Hypertext Preprocessor
PLEs	Personal learning environments
PSLC	Polymer Science Learning Center
SNSs	Social networking sites
3D	Three-dimensional
URL	Unified resource locator
VLE	Virtual |Learning Environment
VR	Virtual Reality
VRML	Virtual Reality Modeling Language
WBE	Web-based education
WBPSS	Web-based performance support systems

WBT	Web-based training
WAN	Wide area networks
WWW	World-Wide-Web
XML	Extensible markup language

Preface

What turns me on about the digital age, what excited me personally, is that you have closed the gap between dreaming and doing. You see, it used to be that if you wanted to make a record of a song, you needed a studio and a producer. Now, you need a laptop.

— Bono

The system of nature, of which man is a part, tends to be self-balancing, self-adjusting, self-cleansing. Not so with technology.

—E. F. Schumacher

The above quotes from U2's lead singer, Bona and E. F. Schumacher capture the paradoxical nature created by the intertwinement of human and machine within the digital age of Web 2.0. Web 2.0 is refers to the "developments in technology employed online that enable interactive capabilities in an environment characterized by user control, freedom and dialogue" (Tuten, 2008, p. 3). The advent of Web 2.0 has dramatically changed the way individuals and organizations interact within society and share information, particularly in regards to educational institutions. As will be demonstrated in this book, the paradoxes within the macrocosm of society are mirrored within our educational institutions where human learning and technology intersect. This is the battlefield where the discussion of e-leaning begins.

Modern e-learning grew up within distance education institutions in the 1970s, led by the United Kingdom and Canada with later participation coming from the United States in the 1980s (see Luppicini [2008] for more complete history). This early growth of e-learning within distance education institutions created a foundation for an e-learning movement that infused traditional public and private systems of educational around the world. This was particularly salient within the corporate sector where flexible corporate training and professional development offered a more efficient educational service (and product) to employees under geographic and time constraints that hampered traditional training and development methods. Now that e-learning was firmly entrenched within education and here to stay, the challenge that continually surfaced was how to make e-learning "good.", how to know if it is "good" when we make it, and what exactly constitutes "good" e-learning?

Consequently, attention to the design and implementation of quality e-learning has spread rapidly within educational institutions over the last 20 years. This is due to a variety of factors including economic advantages, limited building and classroom capacities, increased enrolments, and the ability to bring education to the people by overcoming temporal and geographical barriers that limited access. Obviously, there have been market considerations of a blossoming diploma mill industry, selling a cheap educational product (a course or program) to a limitless group of consumers (students) that spurred the development of much poorly designed e-learning by offering easy to get diplomas to a potential market place that stretched as far as the World-Wide-Web

(WWW) could reach. It was the wild west for educational entrepreneurs that spread like a plague across our digital world during the early popularization stages of e-learning. This has a twofold effect. On the one hand, it created a bad reputation for e-learning that led to public scepticism that created challenges for those of us creating high quality e-learning that was truly advancing education. On the other hand, it drew attention to the need for delivering high quality e-learning to satisfy the needs of an increasingly aware and demanding public.

DIGITAL TECHNOLOGY FOR EDUCATIONAL CONSUMPTION

The indifference of children towards meat is one proof that the taste for meat is unnatural; their preference is for vegetable foods, such as milk, pastry, fruit, etc. Beware of changing this natural taste and making children flesh-eaters, if not for their health's sake, for the sake of their character; for how can one explain away the fact that great meat-eaters are usually fiercer and more cruel than other men; this has been recognised at all times and in all places.

— Rousseau, 1762

Rousseau raises an age-old question about proper education in *Emile* that applies just as well today. Is technology the meat of today, that imposed second nature that our technological society rushes toward, dragging the curriculum and youth along with it? Or, is technology part of the natural experience of youth today, an extension of themselves that they enjoy on their own like milk, pastry, and fruit? And regardless of whether technology is more like meat or pastry, how can educators make it part of a well balanced diet to nurture the youth of today?

What is the mission of educational institutions within our evolving digital world? What has changed and what remains the same? What is the role of educators within an educational system increasingly interwoven with technological advances? And what are the needs of student 2.0 within an evolving digital world where more and more time is spent online. And what kind of education (if any) is suitable for students with the world of information already at their fingertips? It is questions like these that spur educational technologists and educators to push the limits of technology and instruction to provide the best possible education to meet the changing needs of students. As stated by Noroozi and Haghi in this volume, "On-line learning is not about taking a course and putting it on the desktop. It is about a new blend of resources, interactivity, performance support, and structured learning activities" (Chapter 1).

EDUCATING STUDENT 2.0: FACT OR FANTASY

Most learning is not the result of instruction. It is rather the result of unhampered participation in a meaningful setting. Most people learn best by being 'with it,' yet school makes them identify their personal, cognitive growth with elaborate planning and manipulation.

— Ivan Illich, Deschooling Society

Ivan Illich helps allude to the best of what technology brings to education in the twenty-first century—strategic tools that allow educators offer students a meaningful learning experience unhampered by traditional teacher–pupil models of instructional manipulation and unilateral control over the learning process. But like all opportunities, there are risks and challenges to deal with. One of the biggest challenges faced by educators within a technological world is learning how approach pedagogy—how to guide rather than instruct, how to bolster learning rather than constrain it, and how to stimulate individual growth rather than limit it. The other challenge faced by educators within a technological world is to figure out who is being taught to. That is, what are the values and interests of students today, what are their needs and challenges within the current educational climate, and how do we meet them halfway?

Students do not come into the classroom *tabula rasa* (blank slates) with little bobble heads ready for sponge-like absorption within a secure and isolated classroom away from the distractions of the world. This was never more true than today where students carry a variety of information and communication technology (ICT) tools for entertainment and communication with friends and family while in class or on campus. This is the context of classroom life for student 2.0 within our technological world. Student 2.0 has the Internet at their fingertips and the world at their feet. In terms of student perspective, the doors of perception (to burrow from Huxley) are wide open. There are multiple paths to learning rooted in the very nature of the WWW as a distributed network of information. And students are not trapped by it, they eat it and breathe it. To draw on Rousseau's analogy of nature as diet, it is part of their natural diet and the role of educational institutions is to ensure that students get a balanced diet.

To this end, the present volume is divided into three main areas: (1) an examination of e-technology standards and practices within and beyond the walls of educational institutions, (2) a discussion of new approaches to instructional design and teaching that revolve around the use of technology within traditional classrooms and distance education, and (3) an exploration of key experiences and challenges faced by students growing up within a technologically infused educational system that is a microcosm of the larger technology society within which they live. The organization of specific chapters and chapter themes is discussed below.

ORGANIZATION

This volume is put together by a dedicated group of scholars to explore key areas of educational technology research and development within an educational system infused by technology. This system involves administrators, educators, and students, in addition to other stakeholders that care about the advancement of education. To this end it take a multi-faceted approach, pulling together expert work on key areas of technological change that currently affect students, instructors, and educational administrators. Topics covered in this volume include: the development and application of e-learning technology standards, distance learning approaches to online instruction, multimedia classroom practices, perspectives online learning design, Web-based training strategies, and research on students and student experiences with technology within and beyond the classroom learning environment. The organization of this edited volume is as follows:

- In Chapter 1, Noroozi and Haghi explore the importance of e-learning design for engineering education with an emphasis on international e-learning design for use within developing countries. It traces the growing importance that technology plays within engineering education. It then provides a review of leading instructional design models (ISS and ADDIE) applied within engineering education. The review of these leading models highlights e-learning strengths and weaknesses followed by a comparison e-learning and conventional engineering learning methods. The chapter concludes with an insightful discussion of the international potential for designing e-learning with an emphasis on selected developing countries (China, Thailand, Africa, India).
- In Chapter 2, van Hamel and Luppicini provide a systematic research review of social networking sites (SNS) and youth to help provide needed background information on widespread youth activity and interest in SNS. As argued by the authors, "Commercial sites like *MySpace* and *Facebook* may dominate the public imagination when speaking of SNS, but they are merely two examples of the genre. SNS can be repurposed for many goals which serve educational goals." This chapter reviews evidence-based literature on SNS from around the world to summarize what is known about SNS and youth and point out gaps in the research base. The research contributes valuable background knowledge about social networking among youth to help inform education within an increasingly digital world. In addition, this chapter identifies key challenges and opportunities this poses for educators.
- In Chapter 3, Mahdavi and Haghi offer an illuminating discussion of learning design using the Learning Activity Management System (LAMS), a learning design system where collaborative and reflective learning activities can take place. Based on experiences with such systems, the authors review lessons learned and attempt to shed light on possible implications for advancing learning design strategies. The chapter does a great job of articulating key learning design system features in use today. At the same time, the authors acknowledge that the mastery of these systems does not occur over night. As is stated in the chapter, "As much as one might encourage instructors to be creative and transformative, such innovations take time to develop. People often feel comfortable building upon what they are familiar with instead of starting out creative."
- In Chapter 4, Noroozi, and Haghi address the revision of engineering curricula, faculty training, and student needs when considering e-learning within the engineering discipline. For instance, Why not replace a portion of course lectures with online resources and why not offer smaller seminar groups in the time the lectures would have used? The authors review key critical factors that can affect the sustainability of engineering as a discipline, including, learning and teaching methodologies, e-development strategies, contextual considerations, and communications. An emphasis is placed on the potential of e-learning to leverage key aspects of knowledge creation, knowledge sharing, individualized learning, and class participation.

- In Chapter 5, Noroozi, and Haghi explore the contribution of 3D Virtual Reality to e-learning and discuss how 3D resources and models of e-learning can enhance the communication of ideas and stimulate interest of students compared to 2D resources and models. The chapter examines Web3D as a platform for experimenting with new tools and applications for tele-presence and makes a case for 3D Virtual Reality (along with advanced software tools and associated Web technologies) as a means of leveraging e-learning systems by advancing communications and student learning.

- In Chapter 6, Luppicini and Backen contribute to the advancement of teaching practice through the strategic use of technology (blogging and multimedia) and collage inquiry techniques within undergraduate level qualitative research classes. The chapter describes a study on student researcher self-study conducted within a large undergraduate research class offered at a large North American University. A student researcher self-study blog was created where participants posted personal narratives and collage work in response to a series of researcher self-study questions designed by the instructor to promote reflexive learning. The chapter reports on findings obtained from students following this activity and provide insight into the advancement of qualitative research teaching practice. The chapter helps reveal the power of reflective practice within the digital age in helping students explore challenging aspects of their learning and identity. This chapter provides a best practice to aid university instructors and other educational professions.

- In Chapter 7, Noroozi, and Haghi discuss key design and implementation issues in Web-based training (WBT). How can new Web-based learning instructors optimize their use of available tools? How can new Web-based learning instructors organize content into well-crafted teaching systems? This chapter discusses basic requirements for e-learning as a training technique within engineering instruction. It also covers practical considerations which can detract from the success of e-learning. As articulated by the chapter authors, "3D Virtual Reality, software tools, and associated Web technologies are mature enough to be used in conjunction with advanced e-learning systems. 3D based content can enhance communication of ideas and concepts and stimulate the interest of students." The chapter concludes with practical examples on how to apply WBT to selected topics within engineering which should interest various educational professional interested in exploited new technologies for leveraging higher education.

- In Chapter 8, Luppicini and Myhill explore student values within the context of digital education. The chapter is based on the assumption that values shape attitudes, beliefs, and interactions in life and society. Students today have grown up with values emerging from their experiences within an increasing digital world where the use of ICTs are a regular part in school and life. This chapter is framed as a research study on students' core values to find out what they value most in life. To this end, students at a large North American University created a series of artistic and textual descriptions to explore what they value most in life. The chapter discusses study findings revealing the presence of unique attributes aligned with the expression of student values in visual and narrative communications. Implications for educational planning and instructional design are discussed.

- In Chapter 9, Luppicini and Moir present a research study on student ICT use within personal relationships. Undergraduate students at a large North American University created a series of artistic and textual descriptions to explore their personal views on ICTs and their relationships. The chapter examines responses from 63 participants to the question, "How does the use of ICTs affect my approach to relationships?" The data analysis revealed that students found ICTs to be incredibly useful in helping them stay in touch with family and friends across large geographical distances, as well as communicating with professors and other students in their class. The chapter highlights the paradoxical nature of technology within the context of personal relationships. A number of positive and negative aspects of ICT influence on relationships are identified. The chapter provides insight into the powerful influence of ICTs on student relationships and online behavior both within and beyond the university walls. As noted by the chapter authors, "The data gathered from this study helps provide educators with a broader picture of key aspects of student life and relationships in and out of educational settings. Primarily, it provides universities with a glimpse of just how important ICTs are to students. Students need them to gain access to their education, future job prospects, stay in touch with friends and classmates, but most importantly, to stay connected with their family". This chapter provides useful insider information from student stakeholder that educational technologists and instructors can use in developing appropriate technological infrastructure and policies within universities.

- In Chapter 10, Luppicini and Barber take a step back to look at how students navigate their online and offline identities. How do university students compare their online and offline identities? This chapter draws on art-based self-study research methods to examine university students' views of their online and offline identities. Both visual and text-based data were included to capture the multi-faceted nature and complexity of identity representation. Findings revealed that students have a complex approach to online and offline identities that aligned with their conceptualization of online and offline identity as being either the same (continuous) or different. This division in student orientation toward their online and offline identity appeared to influence student online attitudes and behaviors in important ways. This research also found key links between student identity orientation, the influence of anonymity in influencing online behavior, and online identity management strategies among students. Based on chapter findings, the authors insist that, "More educational research should focus on obtaining the 'insider' perspective on online and offline learning from the students affected by it in order to better identify their needs, interests, and barriers."

- In Chapter 11, Luppicini and Lin explore university student views on online anonymity. Although virtual environments are seen as settings for socialization, self-presentation, information exchange, learning, and collaboration, they are also sites for lying, false advertising, deception, and defamation. This is partly due to the anonymity afforded by the virtual environment that allows individuals to be who they want to be and say what they want to say without fear of repercussion or accountability. This creates possible trust issues that may counter efforts to use e-learning to leverage collaborate learning and cultivate online learning communities. This chapter explores how online anonymity influences students' conceptions of themselves and others. Chapter findings shed light on the complex relations between online identity and trust building. The chapter makes the case that, "Educators must integrate media and Internet awareness instruction into the curriculum to ensure that students are aware of the dangers and risks when engaged in anonymous online learning". The chapter helps shed light on the promising opportunity for "an educational re-tooling of the Internet and WWW for learning purposes that is worth exploring, but only with the proper safeguards in place."

- Chapter 12, "Education for a Digital World: Current Challenges and Future Possibilities," Luppicini provides a reflection on the real challenges faced by one educational institution attempting to develop a strategic plan for optimizing program delivery options and technology integration at a large North American University. The chapter addresses some of the practical challenges faced by educational institutions struggling to establish a "shared vision" of educational technology needs. The chapter draws on existing scholarship from the field of Technoethics to help articulate the importance of considering the human side of technology (values and norms) in planning. Under this umbrella, there is an ethical responsibility for educational technology professionals to provide conceptual grounding to clarify the role of technology in relation to those affected by it and to help guide ethical problem-solving and decision-making in areas of activity that rely on technology. The chapter provides practical strategies on optimizing educational technology planning in line with the values of the educational community affected.

Overall, this edited volume provides a diverse collection of chapters that delve into key areas of consideration for education within the digital world. Because the issues related to technology and education are so broad, the *Education for a Digital World: Present Realities and Future Possibilities* is necessarily selective. It attempts to provide in its own modest way a multi-perspective view of key contemporary work on education and technology to help inform educational change within society shaped by and shaping technological developments. Despite the modest aims of this project, the editors acknowledge that it is not possible to please all educational professional, technologists, and general readers. It is hoped that this publication will stimulate ongoing dialog and sharing among those dedicated to optimizing the use of technology in diverse areas of education where there is an established need and interest. This is important to mention since it is all too easy to get swayed by all the "bells and whistles" offered by new technologies. Although the majority of the chapters do deal with various domains of technology and education, there are also serious practical challenges and concerns revolving around technology that educational professionals can gleam from this edited volume. It is the hope of this co-editor that the volume addressed some of the common struggles that many educational professional are faced with when attempting to keep up with technology and re-vision education for a digital world. Critical comments and suggestions are welcome.

REFERENCES

Luppicini, R. (2008). *Trends in educational technology and distance education in Canada.* Berlin: VSM Publishing.

Tuten, T. L. (2008). *Advertising 2.0: Social media marketing in a Web 2.0 world.* Westport, CT: Praeger.

Chapter 1

The Development and Applications of E-Learning Technology Standards

B. Noroozi and A. K. Haghi

INTRODUCTION

Online learning is not about taking a course and putting it on the desktop. It is about a new blend of resources, interactivity, performance support, and structured learning activities. The key benefits of e-learning solutions are: (1) Increased quality and value of learning achieved through greater student access and combination of appropriate supporting content, learner collaboration and interaction, and online support, (2) Increased reach and flexibility enabling learners to engage in the learning process anytime, anyplace, and on a just-in-time basis, (3) Decreased cost of learning delivery, and reduced travel, subsistence costs and time away from the job, and (4) Increased flexibility and ability to respond to evolving business requirements with rapid roll-out of new and organizational-specific learning to a distributed audience. Often an e-learning program will involve development by a number of people simultaneously. Therefore design and development of standards need to be put in place to ensure consistency and transferability of skills. Students are also far less forgiving in terms of inconsistency of user interface and ease of use than with classroom material. The present chapter is a study on the growth of e-learning in engineering education as the most important objective area of science apart from the medical science. Using the definitions and aptitudes of e-learning we tried to find a more specialized way to develop e-learning in engineering education. To this aim, we have investigated the progresses being made on engineering e-learning, and the benefits and difficulties of implementing e-learning in engineering education. The study has focused on the importance of e-learning design for engineering education; therefore, the concept of instructional design has been reviewed and oriented for engineering e-learning. In this regard, application of information systems success (ISS) and ADDIE models parameters in engineering education has been introduced and analyzed. This study also presents an overview on International potential for designing e-learning, with selected case studies of developing countries.

The main issues today in e-learning are: (1) How the learning could encompass large numbers (scalability)? (2) How the learner can gain more insight into the knowledge base interactively? and (3) How easily one can offer the same course on different platforms (Inter-operability)? What security mechanisms are built-in? What level of ease that exists while moving from one tool to another (interchangeability)? Though it is universally accepted that e-learning is the best vehicle for supplementing the knowledge beyond the classroom, it is a point of discussion whether it is more of a supplementary, complementary, or comprehensive learning mechanism. Importance and need of specifications and standards are well known to all of us in different spheres. Right from the building set toys for children, electrical socket and plug, telephone connectors, to serial, parallel bus standards, and many more. Everyone who has ever tried to develop and work with hardware and software understands the important role standards play in facilitating the integration and maintenance of digital system components and data resources. Some e-learning specifications deal with integration. When you buy a building set for your child we expect the child to build different shapes with his imagination using pieces of different shapes. Obviously, the plastic pins and the wholes on these pieces must have accurate dimensions to be fitted together and allow the child to imagine and create new horizons. This seems to be a good analogy to our need of standards in the case of e-learning as well. Designers and developers of online learning materials have variety of software tools at their disposal for creating learning resources. These tools range from presentation software packages to more complex authoring environments. They can be very useful in allowing developers the opportunity to create learning resources that might otherwise require extensive programming skills. Unfortunately, a number of software tools available from a wide variety of vendors produce instructional materials that do not share a common mechanism for finding and using the resources. Thousands of courses are being offered on the net. However, most of the courses offered on the net may not be considered high quality. There is need for standards to define the framework for online learning.

A number of organizations started advocating standards for the learning technology. Standards are desirable for interoperability, convenience, flexibility, and efficiency in the design, delivery, and administration. They provide consistent online dimension for all courses being designed so that all authors/faculty are able to customize the online materials with minimal lead-time. Standards also impose a certain order on chaos resulted due to proprietary products from different vendors. This order provides more uniform and precise access and manipulation to e-learning resources and data. The appropriate standards offer the following benefits:

- Industry-wide standards for learning technology systems architectures;
- Common, interoperable tools used for developing learning systems;
- A rich, searchable library of interoperable, plug-compatible learning content;
- Common methods for locating, accessing, and retrieving learning content; and
- Standardized, portable learner histories that can be transferred with the learner over time.

Recent years have seen dramatic changes in engineering education in terms of increased access to life-long learning, increased choice in areas of study and the personalization of learning. To advance across all domains seems to necessitate incompatible changes to the learning process, as practitioners offer individualized learning to a larger, more diverse engineering student base. To achieve this cost effectively and without overwhelming practitioners requires new approaches to teaching and learning coupled with access to a wide range of resources. Practitioners need to be able to source and share engineering materials, adapt, and contextualize them to suit individual needs, and use them across a variety of engineering educational models (Littlejohn, Falconer, & Mcgill, 2008). Hence, a great deal of effort has focused on the integration of new technologies such as multimedia video, audio, animation, and computers, with associated software, to achieve the improvement of traditional engineering education. The Internet technologies have also been popularly applied to Web-based learning (Hung, Liu, Hung, Ku, & Lin, 2007). The growth of the information society provides a way for fast data access and information exchange all over the world. Computer technologies have been significantly changing the content and practice of engineering education (Gladun, Rogushina, Garcıa-Sanchez, Martınez-Bejar, & Fernandez-Breis, 2008). Information and communication technologies (ICT) are rightly recognized as tools that are radically transforming the process of learning. Universities, institutions, and industries are investing increasing resources to advance researches for providing better and more effective learning solutions (Campanella et al., 2007).

The most important aspects of a focused learning for engineering students are shown in Figure 1.1.

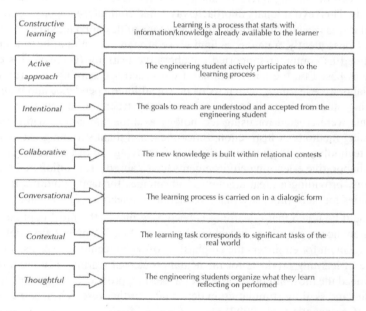

Figure 1.1. The most important and meaningful characteristics for engineering instruction.

The main objective of a learning object is to realize three fundamental learning goals:

1. To inform the engineering students to be responsible for their learning and capable to manage processes to reach aims and to understand their learning needs;

2. To promote real and meaningful learning environments and contexts, enabling the engineering students to retrieve information and build knowledge by using different learning ways; and

3. To create stimulating situations and learning dynamics that prelude to wider learning tasks.

Nowadays, the use of ICT has improved learning, especially when coupled with more learner-centered instruction, or convenience, where learning and exchange with the instructor can take place asynchronously at the learners own pace or on as-needed basis (Motiwalla, 2007). The consequent applications of all multimedia and simulation technologies, computer-mediated communication and communities, and Internet-based support for individual and distance learning have the potential for revolutionary improvements in education (Gladun et al., 2008). For instance, electronic collaboration (e-collaboration) technologies for engineers are technologies that support e-collaboration. An operational definition of e-collaboration is collaboration among different individuals using electronic technologies to accomplish a common task. These e-collaboration technologies include several Internet-based technologies, such as e-mail, forums, chats, and document repositories (Padilla-Melendez, Garrido-Moreno, & Aguila-Obra, 2008).

However, the first computer-supported collaboration system emerged in 1984 from the need of sharing interests among product developers and researchers in diverse fields. This revolutionary approach was called computer-supported collaborative work and it was used to support learning by means of desktop and video conferencing systems. Consequently, a new paradigm arose around educational institutions which were defined as computer-supported collaborative learning (CSCL). This emerging system was based on the contributions of constructivist learning theories about the term collaborative learning, which focus on social interdependence and maintain that engineering students consolidate their learning by teaching one another. CSCL environments were created for using technology as a mediation tool within collaborative learning methods of instruction. Since then, and thanks to the great evolution of network technologies, engineering education is moving out of traditional classrooms. These collaborative e-learning environments have caused a revolution in the academic community, providing a great amount of advantages for using both the Internet and technologies for "any-time, any-place" collaborative learning (Jara et al., 2009).

It should be noted that e-learning is becoming one of the most popular solutions to meet new needs especially in technical courses. In e-learning course development and management for engineers the emphasis is often on technical aspects, whereas the relevance of learning products for the actual process of learning is not considered in depth. Indeed the most important aspect of a learning product is its aptitude to provide knowledge and skill by stimulating in dept study, further researches, and close investigations (Campanella et al., 2007).

After several decades of locally developed, non-portable, and ultimately unsustainable ventures into computer-assisted learning, the e-learning standards and specifications movement has evolved as a way of assuring exchange of educational materials and information between different systems. This is not just a matter of synchronous exchange between different systems, but asynchronous exchange over time so that any one technology can be disposed of and/or replaced while retaining the investment in the content and processes they support.

Of course, there are many other standards underpinning technology use in education: media standards such as USB or CD/DVD-ROM, network standards such as

TCP/IP, and file standards such as text (the basis of all SGMLs children—HTML, XML, and so on). Furthermore, there are many associated meta-standards such as quality assurance of educational veracity and validity, legal standards in the form of laws and licenses (such as Creative Commons), and political standards in the form of policies and procedures (such as section 508 accessibility legislation). For the purposes of this chapter, however, we will limit our scope to the data standards used in support of the interoperability of educational design and intent.

The path to creating a data standard can be a complex one, but in its most basic stages it should be based on addressing a widely held need or problem that can be solved by the standardization of the expression and/or exchange of data. For instance, if many users wish to exchange content, then what should the exchange format be and what should it contain? Alternatively, if users wish to exchange assessment items, then what kinds of question types can be supported? Not only does the process require a bottom-up needs approach, the means of allowing these needs to be expressed, identified, and developed are also required. Professional communities of practice (such as industry or academic consortia) typically form the context for such developments, although requirements for standards may also emerge at the interfaces between communities of practice as a way of allowing them to work together. In this latter context standards can act as "boundary objects" between such communities.

Once a need is clearly articulated and identified then the development process needs a home, preferably within a context that can both sustain the process and be sensitive to its peculiar and critical dynamics. This development context is most likely to be an established standards and specifications development group such as IEEE or LETSI or a domain-specific one such as MedBiquitous. Although there is nothing preventing a unilateral or a non-aligned effort (see Table 1.1 for some of the different educational technical standards and specifications bodies). The result of this initial design process is a specification, usually expressed as a set of formatting rules, typically using extensible markup language (XML). Once the specification has stabilized among those developing it, it is released to the community for testing and implementation. Over time there may be new versions of the specification released to accommodate changing or expanding needs or identified flaws as opportunities to improve on the original design.

Table 1.1. Comparison of e-learning versus conventional engineering learning methods.

E-Learning	Traditional Engineering Learning Methods
It can relies on learners' and it is self-motivation	Lecturer always plays a leading role in motivating and directing the engineering students
Assessment and examinations conduced at learners' place	Assessment and examinations time does not depend on learners
Greater achievement is expected in number of students going through engineering courses	Learner restricted to those attending univeristy or college
Innovative methods required to reach practical assignments and experiments	Laboratories readily available for practical assignments and experiments
Duration of course normally decided by the engineering student	College of engineering has calendars and set durations for courses

A specification is a design statement. For an existing specification to become a standard it must go through an additional formal consultation and validation process by a recognized standards organization such as the IEEE, ANSI, or ISO, at the end of which (assuming it is successful) it becomes a published standard under the publisher's credentials. Although standards can be updated, this happens far less frequently than for specifications—standards are accepted as stable and assured over time. To date, there have been few actual e-learning standards (the IEEEs Standard for Learning Object Metadata is a notable exception) as the somewhat frontier nature of the development and use of educational technologies has meant the number of specifications remains in a state of flux and change. Clearly, despite the civilizing nature of introducing standards and specifications, this is still an environment of discovery and innovation rather than one of limits and constraints (despite the implication that standards and specifications may imply). Nevertheless, there are increasingly common, clear, and consistent design patterns emerging from this environment, the identification and consideration of which can in turn further development and design of both technologies and standards and specifications.

Originating in software engineering, a design pattern is conceived as a general repeatable solution to a commonly occurring problem in software design. A design pattern is not a finished design that can be transformed directly into code. It is a description or template for how to solve a problem that can be used in many different situations. Examples of software design patterns include "proxy; an object representing another object" and composite: a tree structure of simple and composite objects. However, this intrinsically object-oriented metaphor has much wider applications as design patterns are identifiable in many other areas, such as layout grids in graphic design, narrative plot devices (it was all a dream), idiom such as jazz or cubism, and educational designs such as problem-based learning or even the ubiquitous multiple-choice question. The principle of templates and design patterns is not new to educational technology *per se*—they have previously been presented as "general purpose abstract macro-structures."

Perhaps the most common design pattern, not just in educational standards but in almost every area of human activity, is that of the node and the nodal network. The word node has its linguistic origins in the Latin word *nodus*, translated as knot, and can mean many things: a knot or protuberance, a point at which branches divide, a critical or focal point, and a point of intersection or convergence. We see nodal networks in many places—the branches of trees, the rigging of sailing ships, in subway maps, in the structure of circulatory systems in living organisms. And we find it in knowledge models such as mind maps and topic maps, and in databases and wiring diagrams. From the point of view of using nodes as a design pattern it is useful to briefly regard their basis in graph theory, which describes arrays of vertices (nodes) and edges (links) as cyclic graphs with bidirectional edges or acyclic graphs with unidirectional edges.

Traditional education for engineers has shifted toward new methods of teaching and learning through the proliferation of ICT. The continuous advances in technology enable the realization of a more distributed structure of knowledge transfer. This becomes critically important for developing countries that lack the resources and infrastructure for implementing engineering education practices. The two main themes of technology in designing e-learning for engineering education in developing countries focus either on aspects of technological support for traditional methods and localized processes, or on the investigation of how such technologies may assist distance learning. Commonly such efforts are threefold, relating to content delivery, assessment, and provision of feedback. This chapter is based on the authors "10 years" experience in e-learning, and reviews the most important key issues and success factors regarding the design of e-learning for engineering education in developing countries.

E-Learning, Definitions, Applications, and Performance
The impact that Web-based learning can have on users is powerful. The Web is a communication channel with the ability to use it with audio, video, graphics, and text. Users can then communicate through one of these channels and keep in contact with one another, by groups and even in real time. Training programs benefit greatly from the use of the Web for several reasons. These pros consist of training being able to be distributed quickly and easily, graphically dispersed students can communicate and learn effectively, and the update of materials can be a fraction of the coast of them being revised by other means. Some of the things that attribute successful management of a Web-based training (WBT) include an ability to articulate the value that the Web brings to training development, aligning the program to the corporate culture and the effect it has on the organization and also understanding the psychological effects of working with the WBT. The ability to work and manage WBT programs is becoming a norm to most companies. The Internet has influenced almost every aspect of business. With the growth in technology and use of the Internet, training professionals and realizing the plethora of uses that can be applied to existing training programs.

Despite the many benefits of WBT programs, there are drawbacks that must be taken into consideration. Some of the disadvantages may be that fact that online activity may be time consuming, implementation and software updates may be very costly, and the psychology behind WBT may disenfranchise people. Because of these factors, sometimes it is better to continue using traditional training techniques or a combination of both. However, it all boils down to the company and asking themselves if this is something that should invest in. Typical questions to ask may be to what is the nature of the performance deficiency or learning opportunity intended to address, who is the target audience and how will they benefit from it and another may be to ask how will it affect the budget of the company? So with these factors in mind companies need to analyze their current situation thoroughly and weigh out the advantages and disadvantages before commitment.

The technologies of computer graphics have become more and more important, since there are more and more people wanting to use them to create more imaginative content. How, when, and where one can study these useful technologies are good questions for many people, because not everyone majors in computer graphics. With the rapid growth of the Internet, we propose a WBT system and courseware for advanced computer graphics for undergraduate students. By using the WBT system, the students have no temporal and spatial limitations for studying computer graphics technologies.

Once they are able to connect to the Internet, they can study as much as they wish and in any place.

The definition of e-learning for engineers may vary significantly, but perhaps "… learning the engineering concept is aided by information and communication technologies …" is one of the definitions, which is very close to reality. However, some authors and users define e-learning only as "… the delivery of content via all electronic media, including the Internet, intranets, extranets, satellite, broadcast, video, interactive TV, and CD-ROM …" In this case, the emphasis is only at the delivery and many unworthy engineering courses have been "developed" by this way—just delivery to students of existing files with handouts. This has the only advantage of reduced cost, but the educational results do not have significant value. Many authors support the first view that e-learning should develop materials, which will increase the pedagogical effectiveness, and then deliver these to engineering students. Only in this case the full power of e-learning can be utilized (Tabakov, 2008).

E-learning for engineering students, at its best, is the kind of learning that complements traditional methods and gives a more effective experience to the learner (Magoha & Andrew, 2004). E-learning refers to the use of electronic devices for learning, including the delivery of content via electronic media such as Internet/Intranet/Extranet, audio or video tape, satellite broadcast, interactive TV, CD-ROM, and so on (Shee & Wang, 2008). Simply, e-learning for engineers is the use of technology to support the learning process. Fundamentally, it is about putting the learner first by placing resources at the learner's fingertips. The engineering e-learner is able to dictate the pace and balance of learning activities in a way that suits him/her. E-learners can absorb and develop knowledge and skills in an environment that has been tailored to suit them—and at their own pace. As opposed to online courses in their strictest sense, e-learning does not necessarily lead to an engineering certification or an engineering degree programmer but may be tailored, for example, to suit the needs of a specific company (Magoha & Andrew, 2004).

It should be noted that technology has proved its value in engineering education and in applied areas such as engineering management. For example, in Europe digital literacy is emerging as a new key competence required by workers and citizens in the new knowledge society. The integration of IT-supported learning helps workers acquires the necessary skills and knowledge for their job. IT use can also improve the effectiveness of the learning process. Consequently, if the engineering student learns to use technology before starting his/her job; this could be an advantage for both the future profession and the employer. Moreover, the use of the Web as an educational delivery medium (e-learning) provides engineering students with the opportunity to develop an additional set of communication, technical, teamwork, and interpersonal skills that mirror the business environment in which they will work. Meanwhile, the statistics on e-learning show a considerable use of these tools in recent years. Universities are combining interactive technology and active ways of learning, which require students to develop or hone their computing skills and to take more responsibility for their own learning. Nevertheless, engineering students, contrary to the general idea that they can be considered digital natives, do not all react positively to IT learning;

some prefer the traditional process. Engineering students may react differently to the online learning environment, depending on their own level and attitude. This is similar to the finding about teachers working in public centers who have shown some resistance toward IT implementation in engineering education. There is also the need to investigate the engineering students' acceptance of an Internet-based learning medium in order to understand the various drivers influencing acceptance (Padilla-Melendez et al., 2008).

E-learning for engineering students is important for building a technologically literate workforce as well as for meeting societies continuous need for rapid life-long learning delivered in increasingly more convenient forms (Buzzetto-More, 2008). In spite of all efforts, in the last years e-learning has experienced slow user growth involvement and high dropout rates in many organizations: users become easily frustrated or unenthusiastic about the material and do not complete learning activities (Campanella et al., 2007).The development of e-learning materials can be presented as a multi-layered process, including the following stages:

- Programming specific simulations;
- Building of e-learning modules; and
- Development of e-learning programs.

These stages most often exist as separate entities, but the programs will include modules, and simulations can be applied at various stages (Tabakov, 2008) to provide five approaches for using technology in learning (Motiwalla, 2007):

1. Intelligent tutoring systems that have attempted to replace the teacher; but these have never been successful due to their limited knowledge domains;
2. Simulation and modeling tools that serve as learner's assistants or pedagogical agents embedded in applications that act as mentors providing advice;
3. Dictionaries, concept maps, learning organizers, planners, and other resource aids that help learners to learn or organize knowledge with system tools and resources;
4. Personalized communication aids that can present materials depending on user abilities and experience with the system; and
5. Simulated classrooms and labs that engage teachers and learners in an interaction similar to the real classrooms.

The effect on e-learning is measured with an information systems success (ISS) model because it is also one of the information systems. The e-learning success model evaluates e-learning effectiveness based on the ISS model, constructivism, and self-regulatory efficacy. In 1992 an ISS model has been suggested that is measured through six dimensions as is presented in Figure 1.2.

System quality implies an information process system quality based on production of produced information. Information quality is defined as the quality of system product outputs, and the usage and user satisfaction is defined as the recipients' interaction of information and information system product. Also, individual and organizational effects are measured by the information system which affects the users and the affiliated organization of the users (Lee & Lee, 2008).

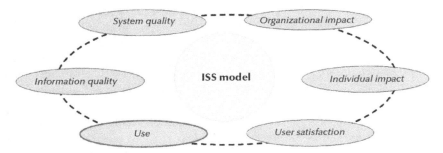

Figure 1.2. ISS model parameters.

E-Learning for Engineering Education

Despite the dramatic expansion in e-learning and distance education, e-learning in engineering education still faces a number of setbacks that prevent an equivalent expansion rate. For effective and complete learning in engineering, engineering education requires a mixture of theoretical and practical sessions. In order to understand how theoretical knowledge can apply to real world problems, practical exercises are essential. While it is relatively easy to simulate experiments, performing practical experiments online has continued to be a challenge. Coupled with this, engineering software is often very expensive and may not be easily affordable by the ordinary e-learner. Although low-cost alternatives that utilize freeware have been successfully developed and tested, practical laboratories that support engineering education are still difficult to implement online (Magoha & Andrew, 2004).

Engineering education relies heavily on capital-intensive laboratory equipment. Collaboration with developed countries would, therefore, be one path to enhance learning in engineering concepts for developing countries. A collaborative laboratory component can bridge the gap between regular e-learning and e-learning in engineering. This can be achieved through an Internet laboratory, examples of which are several Internet-based laboratories have already been developed and have been described elaborately. Most of these laboratories are based on standard hardware systems, such as those provided by National Instruments (NI), and are commercially available. However they are costly and often require specialized training. The solution is to develop low-cost systems that rely partially on expensive hardware at the server and open software at the client (Magoha & Andrew, 2004).

Numerous digital applications have been developed to demonstrate laboratory experiments, real-world landscape processes, or microscopic objects. Engineering students learned faster and liked their classes more than students in traditional on-campus classes. Cognitive scientists believe that to learn, the material must have meaning to the learner. Cognitive science may be defined as a multi-disciplinary approach to studying how mental representations enable an organism to produce adaptive behavior and cognition. The following criteria have been considered for certain virtual field laboratory (Ramasundaram, Grunwald, Mangeot, Comerford, & Bliss, 2005):

- Global access, that is, Web-based implementation;
- Stimulation of a variety of learning mechanisms;

- Interactivity to engage engineering students;
  - Compartmentalization and hierarchical organizational structure; and
- Abstraction of 2D and 3D geographic objects (e.g., soils, terrain) and dynamic ecosystem processes (e.g., water flow) using geostatistics and scientific visualization techniques.

Medical Physics and Engineering (MEP) is another example among the first professions to develop and apply e-learning. An indicator for this is the first international prize in the field (EU Leonardo da Vinci Award) presented to European Medical Imaging Technology (EMIT) Consortium in 2004. During the last 15 years, a number of activities and publications addressed the questions of MEP Education and Training. These led to rapid development of the profession worldwide and now the next stage of e-learning is to be addressed. A special issue on the subject was published by the Journal of Medical Engineering and Physics in 2005. Based on the paper and on the authors' 12 years' experience in e-learning, the following key elements can be specified for the introduction and use of e-learning in MEP:

- E-learning is imperative for engineers because it offers quick and easy update of teaching materials—a very important function for this dynamic profession. This, combined with the fast delivery of the content through Internet, makes e-learning materials the first choice for many engineering lecturers.
- E-learning proposes an elegant way to solve the engineering problem through the understanding of complex physics models. Using interactive simulations, computer diagrams or images leads to increased effectiveness of the engineering learning process.
- Images are specifically important for engineers and e-learning provides a cheap and effective means for publishing large number of images (either on CD or through the Web). Additionally e-learning can offer a means of image manipulation, which has no analog in other means of publications.
- The search function offered by various e-learning materials is another important advantage. This is also imperative for engineers.
- Finally, the fact that many engineering students from around the world can use the materials through the guidance of most renowned specialists has no analog in the other educational methods and media (Tabakov, 2008).

E-learning Benefits and Difficulties for Engineering Education
E-learning system for engineering education is able to be interpreted in various ways such as "computer based, education delivery system which is provided through the Internet," or "an educational method that is able to provide opportunities for the needed people, at the right place, with the right contents, and the right time" (Lee & Lee, 2008, p. 33).

Some difficulties have been noted in using technology for professional development and four suggestions offered for product refinement and use. The learner must: (1) understand the navigation system, (2) spend time focusing on domain knowledge, rather than on navigation, (3) be prepared in domain area knowledge [e.g., mathematics] or be experienced in learning in technology environments, or (4) all three aspects. These attributes and concerns suggest the need for continued investigations about outcomes of the use of these programs in education and discussions about further product design (Pryor & Bitter, 2008).

E-learning for engineering education offers unique pedagogical opportunities to enhance student learning: In the realm of e-education, there are clear benefits that can be derived from e-learning as follows:

- It promotes exploratory and interactive modes of inquiry.
- It supports and facilitates team-orientated collaborations and expands the ease of access to engineering education across institutional, geographical, and cultural boundaries, among others.
- Class notes and materials are posted on the Internet and students can access the sites from anywhere in the world.
- This is quite unlike distance learning, where an engineering student is given course materials and reads solely on his/her own until examination time.
- E-learning is interactive; the software permits the engineering student to communicate, not only with the lecturer, but also with fellow classmates. It enriches and supplements the classroom experience by engaging the Web.
- E-learning has the ability to communicate consistently to learners by providing the same concepts and engineering information—unlike classroom learning, where different instructors may not follow the same curriculum or teach different things within the curriculum.
- E-learning is cost effective in terms of learners per instructor. In addition, it saves classroom time and this is very significant for learners who are employed on a fulltime basis.
- Engineering students, instructors, and evaluators can track learning outcomes more easily (Magoha & Andrew, 2004).

It has been found that all of the respondents considered the online course materials beneficial to their overall learning experience. Obtained results have shown a positive effect on engineering student learning, problem-solving skills, and critical thinking skills, with females responding more positively than males. Findigs revealed that students who took the quizzes online significantly outperformed students who took the pencil-and-paper quizzes. It has been asserted that Web-enhanced learning improves instruction and course management and offers numerous pedagogical benefits for learners. Engineering students in Web-enabled learning environments become more active and self directed learners, who are exposed to enhanced learning materials. Course websites have proved to be an effective means of delivering learning materials, with students responding positively to the quality resources they make available (Buzzetto-More, 2008).

Although the significant advantage of digital resources is that they can offer flexibility of format, and ease of storage and retrieval, however, digitization, on its own, is not enough to ensure flexibility and use. A major difficulty with early digital "multimedia" resources was their inflexibility; they were not designed for reuse across a range of contexts. Consequently, resources were often based around a single educational model and made available in a set format. Ease of finding suitable resources is another important factor in their effectiveness, as suggested by studies showing that the current generation of students often chooses to source digital resources in preference to print-based materials and their habitual use of "Google" as a primary resource search engine. Among the reasons given for these preferences are that the computer terminal provides a "one stop shop" for resources, and that while *Google* may not provide the best quality information or most efficient search, it is familiar and has a track record of producing results that are adequate. Similar criteria of unified access, familiarity, and adequacy seem likely to apply to teachers' strategies in sourcing resources (Littlejohn et al., 2008).

For all its potential in dealing with learners individually, online learning does have its own invisibility problems. The other side of online learning can seem far away and isolated. For the learner, the price of empowerment is the responsibility of wielding that power. As with all distance learning, e-learning relies on self-motivation. With no enforced discipline or deadlines, it is easy for the learner to be distracted and put off work for a distant tomorrow. With no human presence, it is also impossible for a learner with a problem to obtain help easily. Thus, personal interaction between the instructor and engineering students is either absent or else very different from traditional face-to-face learning. Other subtle disadvantages of e-learning include the ability to read text from the computer screen. Research has shown that linear text is often difficult for people to read from the computer. Hence, learners often have to reformat the text and print it out for reading, necessitating the need for a printer. The problems of invisibility, anonymity, and isolation can be dealt with in several ways. Once more, the key solution is communication and there are several channels available to learners. The first is a national help line where learners can talk to someone trained to give them advice on their course. Many e-learners are also assigned an online tutor who checks on their progress and can be contacted by e-mail with any queries or worries. Online message boards and chat rooms provide learners with the means to contact and talk to other people doing the same course, exchanging advice, ideas, and encouragement. For those without access to computer at home or who require additional support there is a network of centers across countries. Surely, no engineering educator would argue against a legitimate teaching method that highly motivates students to learn their subject matter and promotes student interaction in the process. To witness students achieve an increased mastery of ideas, while also setting greater expectations for themselves, would be fulfilling educators' dreams. Table 1.1 summarizes a comparison between e-learning and traditional learning (Magoha & Andrew, 2004).

Instructional Design and Engineering Education
We believe that the design and implementation of e-learning functionalities in the form of Web services provides mechanisms to reuse e-learning functionalities and to create flexible platforms that can easily be adapted to the individual needs of learners. By application of already existing standards from the field of Web services, functionalities can easily be integrated. It should be noted that Instructional Design (ID) is a multi-disciplinary domain that helps people learns better. ID has unwavering focus on

learning engineering goals, content types, learners, and technology. Therefore, the use of ID in e-learning ensures effective learning design.

Our past experiences show that ID is the systematic development of instructional specifications using learning and instructional theory to ensure the quality of instruction. It is the entire process of analysis of learning needs and engineering goals and the development of a delivery system to meet those needs. It includes the development of instructional materials and activities, testing, and evaluation of all instruction and learner activities. As a discipline, ID is that branch of knowledge concerned with engineering research and theory about instructional strategies and the process for developing and implementing those strategies. Nevertheless, this is the science of creating detailed specifications for the development, implementation, evaluation, and maintenance of situations that facilitate the learning of both large and small units of subject matter at all levels of complexity (Siemens, 2002). Based on our teaching experiences, implementation of this multi-disciplinary science in e-learning and its design, the design of e-learning for engineering courses seem necessary. Hence, in this session we try to make a clear scheme of ID and relate it to engineering e-learning.

There are different models for designing instruction, as there are different learning theories. Martin Ryder (2008) of Denver School of Education at University of Colorado has divided the ID models into two groups:

1. Modern models (including several models in three categories: Behaviorist, Cognitivist, and Prescriptive Models); and
2. Postmodern phenomenological models (including several other models in Constructivist Models category).

Among all presented models in ID, ADDIE which is a prescriptive design model has been selected here to be developed according to engineering e-learning demands. ADDIE is one of the first design models. There has been much discussion about its effectiveness and appropriateness. ADDIE has been selected for its simplicity, ease of application, and cyclic nature.

In 1975, Florida State University developed the ADDIE five-step model of (see Figure 1.3):

1. Analysis (Assess and analyze needs)
2. Design (Develop materials)
3. Development (Develop materials)
4. Implementation (Implement activities and courses)
5. Evaluation (Evaluate participant progress and instructional materials effectiveness)

ADDIE was selected by the Armed Services as the primary means for developing training. At the time, the term "ADDIE" was not used, but rather "SAT" (Systems Approach to Training) or "ISD" (Instructional Systems Development). As a general rule, the military used SAT, while their civilian counterparts used ISD. The "D" in "ISD" first stood for "Development" but now normally means "Design"(Boot, van Merrienboer, & Theunissen, 2008; Clark, 2004; Dick & Carey, 1996).

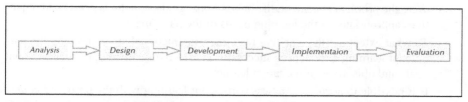

Figure 1.3. Five-steps of ADDIE model.

Analysis (*Assess*)

In analysis as the first step of ADDIE, the designer identifies the learning difficulties, the aims and objectives, the learners' needs, knowledge, and any other relevant characteristics. Analysis also considers the learning situation, any constraints, the delivery options, and the timeline for the project (Colston, 2008). It should be noted that in engineering education, this concept needs very special attention.

Design

Design is concerned with subject matter analysis, lesson planning, and media selection. The choice of media is determined by contingencies of the participant's needs and available resources. This is a systematic process of specifying learning objectives. Detailed storyboards and prototypes are often made, and the look and feel, graphic design, user-interface, and content are determined here (Colston, 2008). Malachowski (2002) from city college of San Francisco, believes that lesson planning requires that instructors determine:

- Objectives defined in terms of specific measurable objectives or learning outcomes;
- Skills, knowledge, and attitudes to be developed;
- Resources and strategies to be utilized;
- Structuring, sequencing, presentation, and reinforcement of the content; and
- Assessment methods matched to the learning objectives to ensure agreement between intended outcomes and assessment measurements.

Joyce and Flowers list seven instructional functions to show how to incorporate available technology into presentations, namely, informing the learner of the aims, presenting motivations, increasing learner interest, helping the learner review what they have previously learned, providing conditions that will irritate presentation, determining order of learning, and prompting and guiding the training (Colston, 2008; Malachowski, 2002).

Development

The actual creation (production) of the content and learning materials is based on the design phase. Development is a process of creation and testing of learning experiences where one tries to seek and answer questions such as:

- Have the learning needs and characteristics of the participants been accurately analyzed?

- Were the problem statement, the instructional goals, and the instructional objectives appropriate for the learning needs of the participants?
- To what extent are the teaching resources, instructional strategies, and the participant learning experiences successful in effectively meeting the instructional goals and objectives of the target learner?
- Is it possible to accurately assess participant learning with the proposed course of instruction? (Colston, 2008; Malachowski, 2002).

Unlike other branches of educations, e-learning in engineering education is a highly complicated concept. Here are some important points to which the designer has to pay special attention:

- Boring and weak instructional content;
- A technology where its application is difficult; and
- A culture which is not informed of e-learning or developing sorts of e-learning designs which are not consistent with the culture. This point resonates more in developing countries or countries with fewer Internet users.

Implementation

Negative responses indicate a need for revision. Implementation is the presentation of the learning experiences to the participants utilizing the appropriate media. Learning, skills, or understanding, are "demonstrated" to the participants, who practice initially in a "safe" setting and then in the targeted workspace. It may involve showing participants how to make the best use of interactive learning materials, presenting classroom instruction, or coordinating and managing a distance-learning program. The progress of the learning frequently follows cyclic patterns based on motivation and intention. Curriculum should be organized in a spiral manner such that the participant continually builds upon what they have already learned. During implementation, the plan is put into action and a procedure for training the learner and teacher is developed. Materials are delivered or distributed to the student group. After delivery, the effectiveness of the training materials is evaluated.

Evaluation

One purpose of evaluation is to determine if the instructional methods and materials are accomplishing the established goals and objectives. Implementation of instruction represents the first real test of what has been developed. Try to pre-test instruction on a small scale prior to implementation. If this is not possible, the first actual use will also serve as the "field test" for determining effectiveness. This step consists of two objectives (Willis, 2008):

1. *Develop an evaluation strategy*: Plan how and when to evaluate the effectiveness of the instruction.
2. *Formative evaluation* can be used to revise instruction as the course is being developed and implemented. For example, the distance educator can give students pre-addressed and stamped postcards to complete and mail after each session. These "mini-evaluations" might focus on course strengths and

weaknesses, technical or delivery concerns, and content areas in need of further coverage.

Within the context of formative and summative evaluation, data are collected through quantitative and qualitative methods. *Quantitative evaluation* relies on a breadth of response and is patterned after experimental research focused on the collection and manipulation of statistically relevant quantities of data. In contrast, *qualitative evaluation* focuses on a depth of response, using more subjective methods such as interviews and observation to query a smaller number of respondents in greater depth. Qualitative approaches may be of special value because the diversity of distant learners may defy relevant statistical stratification and analysis. The best approach often combines quantitative measurement of student performance with open-ended interviewing and non-participant observation to collect and assess information about attitudes toward the course's effectiveness and the delivery technology (Willis, 2008).

Summative evaluation consists of tests designed for criterion-related referenced items and providing opportunities for feedback from the users. Revisions are made as necessary (Colston, 2008). This is conducted after instruction is completed and provides a database for course revision and future planning. Following course completion, consider a summative evaluation session in which students informally brainstorm ways to improve the course. Consider having a local facilitator run the evaluation session to encourage a more open discussion (Willis, 2008). Following implementation of the course/materials, collect the evaluation data. Careful analysis of these results will identify gaps or weaknesses in the instructional process. It is equally important to identify strengths and successes. Results of the evaluation analysis will provide a "springboard" from which to develop the revision plan (Willis, 2008). The benefits of evaluation in engineering e-learning are: (1) Justify and expand training funds, (2) Selecting the best instruction strategies, (3) Obtain and analyze learners' fulfillment feedback, and (4) Quantify performance improvement. And the risks of not evaluating are: (1) Waste money on ineffective programs, (2) Terminate training program, (3) Ineffective and inefficient transfer of training, (4) Training program does not support objectives, (5) Learners' training needs not met or unknown, (6) Presentation improvement unknown, and (7) Expectations are not set nor communicated (Astrom, 2008).

International Potential for Designing E-learning: Selected Case Studies in Developing Countries

There are numerous e-learning programs in developing countries. In Africa, for example, several distance-learning projects have been implemented successfully, the majority through collaboration with institutions in developed countries; a comprehensive list can be found from the African Distance Learning Association. Concerted efforts through outside collaborations and partnerships with developing countries and a number of African universities have culminated in various distance learning degree programs. The African Virtual University offers a range of degree programs, most of them in collaboration with Australian institutions. For developing nations, the realization of e-learning, especially in engineering education, is yet to be thought of. There are several obstacles that developing countries have to overcome if e-learning is to be successfully implemented. The first and major obstacle is the lack of technological resources. While the Internet has been around for over 15 years, it is only in the last few years that most developed countries have had access or even had networks on the Internet. While the Internet is dependent upon a reliable telecommunications system, most such networks in developing countries are badly maintained and often do not have the requisite bandwidth to support e-learning activities. In many developing countries, such networks belong to government agencies and are usually underfunded. A second obstacle is the lack of skilled personnel who are capable of implementing and maintaining an e-learning environment. According to ILO statistics, for instance, less than 1% of employed workers in Africa are professionals with an IT background. All these factors, together with the lack of financial resources, have contributed to the slow growth in e-learning in engineering education in developing countries to date. While the progress is slow, a framework for sustained growth could be created through collaborative efforts, especially among institutions in developing countries and partnerships with institutions in developed nations (Magoha & Andrew, 2004). The difficulties of e-learning evolution in developing countries can be counted as:

- Lack of appropriate technological substructures;
- Financial problems;
- Lack of awareness on IT and information on the benefits of e-learning;
- Problems in the way of communication, which are primarily the language skills, or computer skills; and
- Cultural consistency.

However the developing countries have potential propensities motivating the growth of e-learning due to the growing demand for knowledge and technology, the easiness and cost-effectiveness comparing to following the education in a developed country, and encountering with living costs and communication difficulties in these countries, and the expanding growth of Internet utilization. In the following sections a brief review on the efforts and accomplishments of some Asian countries on e-learning for engineers is presented.

E-learning Progress for Engineers in Thailand: "A Brief Review"

E-learning in Thailand like many other developing countries, deals with barriers such as culture, organization, limited technology, and availability low household penetration of the ICT such as telephones, satellites, PCs, Internet connections, broadband, and so forth. Again, technologies available during the last decade did not appear to support the engineering education sector perfectly. They have been rare, limited, and relatively expensive for engineering education purposes. Multimedia presentation is a well-established (and well tested) form of learning that requires a large bandwidth for the transmission of the contents to all destination users, certainly pushing the broadband diffusion too. It is important not to forget the localization as Thai students do not love studying non-Thai contents and this is why almost well known foreign (English) contents did not succeed in Thailand market. The majority of existing e-learning contents available are in English. Thus, e-learning contents in non-English are highly in demand (Sirinaruemitr, 2004; Vate-U-Lan, 2007; Witsawakiti, Suchato, & Punyabukkana, 2006).

The first traces of e-learning for engineers in Thailand returns to 1993. When after several years of usage of Internet in academic centers in Thailand (i.e., Asian Institute of Technology and Chulalongkorn University), they felt that Internet should be made available to the whole Assumption University (AU). The outcome was a project called AuNet with the following objectives:

- To educate the engineering students, faculty, and staff members on the concepts of local and international networking;
- To prepare the engineering students to enter into information society where networking will be the norm rather than the exception; and
- To provide full Internet access to all engineering students, faculty, and staff members for their personal and educational usage (Chorpothong & Charmonman, 2004).

In response to the rapid expansion of knowledge and widespread demand of engineering education in 2002, this project on e-learning was proposed with the following aims:

- To serve the country by allowing those interested in engineering education the opportunity to continue their studies conveniently, no matter from where or when;
- To promote life-long learning by using the Internet;
- To expand AU from traditional classroom-based engineering education to Internet-based distance education; and
- To increase the number of students at AU from about 18,000 persons in 2002 to 100,000 persons later (Chorpothong & Charmonman, 2004).

The establishment of the College of Internet Distance Education on April 25, 2002, with over 2000 Internet terminal and 1000 phone lines, and mirror sites for degree programs from other countries was the first step of this project which focuses on degree-level e-learning in Thailand. The College of Internet Distance Education of AU has entered into cooperation with many universities such as UKeU, Middlesex, and UNITAR, hoping that by the year 2010, the target of 100,000 students, will be reached. Now the attempts on developing e-learning seem to be more on e-learning designing and overcoming infrastructure problems. Local case studies are more emphasizing on the role of design of hypertext or hyperlink and the flexibility of presentation (Morse & Suktrisul, 2006; Siritongthaworn & Krairit, 2004; Vate-U-Lan, 2007; Witsawakiti et al., 2006).

Furthermore, Thailand is quite slow in deploying the services including the necessary (broadband) infrastructure and transforming the ways of learning and the (digital) contents are very rare too. In some open universities such as Ramkhamhaeng University, and Sukhothai Thammathirat University, it's true that they are working on e-learning projects but they are still familiar with one way broadcasting or two-way video conferencing through the satellite and fiber optics. This is because the digital contents available may not deliver to the learners or students by means of using effective broadband ADSL for lack of availability. As a step ahead, ICT ministry of Thailand has launched a program to address provincial home market and enabled people in the rural areas to have home computers. In addition, low-cost software has been introduced to support the hardware program and in turn Thailand can now see quite a low cost of Microsoft licenses and freeware such as Linux and Pladao being deployed. Ministry sponsored low-cost software's and cheap computers as a result of high competitions and reduced production cost have for the first time allowed Thai people to be able to afford reasonable computers at home in larger and growing quantities. However, people still perceive that Internet access cost is still too high. This is the barrier to the use of Internet for education. Broadband access fees are also still relatively high and are considered as a premium. Affordability is still an important consideration as far as broadband was concerned (Sirinaruemitr, 2004).

E-learning Progress for Engineers in India: "A Brief Review"

E-learning in India is gaining prominence slowly, but indeed steadily. This is due to the fact that more than half the population of India today is below 25 years of age and the number of Internet users is growing continuously. According to UNESCO reports at 2006, 65.2% of adults and 81.3% of youth are literate (UNESCO, 2006a). The tremendous growth of the economy in the recent past has also helped in the growth of online education in India. E-learning in India is especially popular with the young professionals who have joined the work force quite early but still would like to continue their education that may help them move up their career ladder quickly and safely. They find online education in India very convenient, as the nature of the course work does not require them to attend regular classes. Moreover, reputed institutes like Indian Institute of Management, Indian Institute of Technology, and Indian Institute of Foreign Trade are today offering e-learning courses. The scope of online education in India is actually much wider. Apart from proper course works, some e-learning portals in India are also conducting mock tests for various competitive examinations like engineering, medical, management, and so forth.

In India, the engineering education has got both government and private players in the market. It consists of arts, science, and management, technical and professional education. Since the Indian knowledge industry is entering into the take off stage, the strategy of survival is high. The foreign players are also trying to join the competition. And hence the less effective educational institutions for engineers are forced to merge themselves with others or they are forced to go out of market. However, there is a paradox in e-learning among various institutes in India. Few institutions join the race, while the rest suffer from lack of knowledge or from lack of realization of the importance of e-learning. Institutes like IITs are adopting all latest technologies and are keeping their engineering students enlightened from various parts of the world. E-learning has vast potential in India.

The IT service sector export of India has grown up from US $754 million in 1995–1996 to US $ 12,000 million in 2004–2005. The annual growth based on the trend analysis is US $573 in India and US $1,038 in Tamil Nadu. The Indian IT sector especially Tamil Nadu IT sector is growing at a faster rate. This same speed of growth is not replicated in e-learning application. If all these efforts, are directed properly the ups and downs in knowledge growth can be removed.

In engineering education, virtual classroom or a teacher free classroom has got bright future in India. A virtual classroom is one where the virtual reality is enhanced. It is a totally technology-enabled class room environment. Virtual Reality (VR) is a 3D learning environment where the learner can explore the learning concept. Learning is experienced through games or simulated situation. It brings a real environment while wearing a headset and data glove in an immersive virtual reality environment. In the areas of Medicine, Engineering, Astronomy, and other skill trainings, virtual e-learning will become indispensable (Nelasco, Arputtharaj, & Alwinson, 2007).

E-learning Progress for Engineers in China: "A Brief Review"
Cyber-education has been growing rapidly in China since the late 1990s, especially in the fields of higher education and basic education. Technical and engineering institutions in China are currently encountering a high pressure onto their schooling capacity due to an increasing population of education pursuers. Cyber-education is considered as a fast and economic approach to ease the pressure. In September 1998, special licenses were granted to Tsinghua University, Beijing Post and Telecommunication University, Zhejiang University, and Hunan University as the first set of higher educational institutions pioneering cyber-education. In 1999, Beijing University and the Central Broadcast and Television University were added to the pioneer list. Simulated by favorable policies, a considerable number of Chinese universities started to invest in cyber-education since then. At the end of 2002, up to 67 universities in China have received cyber-education licenses. It is estimated that over 1.6 millions of students are enrolled in these cyber-education institutions, involved in 140 specialties from 10 academic disciplines. Besides, the Central TV University which has over 2 million of students, enrollment is moving to cyber-education (Zhiting, 2004).

E-learning Progress for Engineers in Iran: "A Brief Review"
The first traces of e-learning returns to 25 years ago with distance learning. Iran is a country with a high interest in education. This country has an expanded network of private, public, and state affiliated universities offering degrees in higher education. Reports of UNESCO show that up to 2006, 84.0% of adults and 97.6% of youth are literate and this is well balanced between male and female population. Furthermore, a significant majority of the youth population is at or approaching collegiate levels (UNESCO, 2006b). However, the tendency of getting higher levels graduates is obviously growing between the youth population and particularly in graduated students.

Several universities in Iran have started e-learning programs for distance education, such as Sharif University of Technology, the most prestigious technical university in Iran and Middle East. It has online courses on mechanic, electronic, and advanced modeling methods. However, many other universities such as University of Tehran, Amirkabir University of Technology, University of Science and Industry, and University of Shiraz have developed several online courses with the main concentration on engineering education. Despite of all the efforts being made to expand e-learning methods in educational structure of Iran, the process of transferring the traditional education in the Iranian society involves many pressing difficulties which according to the recent studies can be summarized as:

- Lack of realistic comprehension concerning the process of learning;
- Ambiguous understanding about students' educational needs in different levels;
- Defective implementation of computer hardware and software;
- Limited experienced IT professionals; and
- Incompatible educational resources for e-leaning (Dilmaghani, 2003; Giveki, 2003; Hejazi, 2007; Noori, 2003).

DISCUSSION

This chapter weighed the potential challenges and benefits of implementing e-learning for engineers in developing countries. It provides a broad perspective or e-learning design in engineering education. The main themes of technology in engineering education for developing countries focused either on aspects of technological support for traditional methods and localized processes, or on the investigation of how such technologies may assist e-learning design. Key efforts relate to content delivery, assessment, and the provision of feedback. This chapter also reviewed key issues regarding the implementation of e-learning for engineering students in developing countries that could be tailored to satisfy the needs of a limited educational infrastructure. Moreover, for e-learning design to succeed in the developing world, it needs to build one important pillar: the existence of infrastructure, along with some degree of connectivity. Nevertheless, a key challenge revolves around technological requirements which must be kept to a minimum in order to increase the participation of developing countries in e-learning design for engineering students.

The development of technical standards and specifications shares many design and development issues with other information systems in that they all benefit from parsimony, requisite complexity, usability, understandability, and applicability. However, there are some extra dimensions that introduce particular challenges for standards and specifications designers and implementers alike. One of the most important of these is that standards and specifications need to be abstracted from many instances with the purpose of rendering them as relevant to as wide a range of instances as possible. Furthermore, while standards and specifications may for some simply function as a transport mechanism for "what goes over the wire" between different systems, for others they can act as templates for developing whole information systems, encapsulating as they do (or at least they should) the collected best thinking and intent of experts in the field. In this latter context design patterns take on an increased level of significance both as "a general repeatable solution to a commonly occurring problem" and as a template or idealized design for other systems and technologies that employ it.

The development of "education informatics" represents a more holistic and integrated view of the role of information systems and related practices to all aspects of the education environment. The education informatics conceptual space includes business and enterprise approaches in support of logistics and administration, systems to support teaching, learning and assessment, content and knowledge management, and meta-systems such as curriculum maps and outcome frameworks. The development of standards and specifications is a key aspect of education informatics, both in terms of their ability to connect and integrate across the education informatics domain, and the ways they draw out and reify common patterns and approaches for all to see and employ. The approach outlined here can be abstracted further to a series of (at least provisional) education informatics design pattern principles:

- Relational and procedural structures can be modeled using nodal frameworks.
- Simple primitive structures can be instantiated and aggregated in different ways to create more complex structures.
- Specific designs can be made more generic by moving the specificity to semantic markup so that structure and semantics are separate.
- Elements and collections of elements should be uniquely addressable using resolvable unique IDs.
- Non-dependent and out-of-scope functions can be cast as abstract (non-specified black box) services.
- Stateless and state-dependent phenomena need to be identified, handled, and modeled as fits their form and function.
- Non-formal systems can be modeled by identifying common and recurring patterns.

KEYWORDS

- **Electronic collaboration**
- **Electronic learning**
- **Engineering education**
- **Information and communication technologies**
- **Web-based training**

Chapter 2

Beyond the Classroom Walls: A Review of the Evidence on Social Networking Sites and Youth

Anton van Hamel and Rocci Luppicini

INTRODUCTION

In the public imagination, social networking sites (SNS) have been branded as a source of problems ranging from teenaged narcissism to replacing face time with close friends to abduction by sexual predators. Yet despite the impulse on part of parents, educators and policymakers to curb these risks, there is comparatively little empirical research on how SNS is incorporated into the everyday lives of those 25 and under, and the early findings are far from frightening. This research review compiles the extant, evidence-based literature on SNS from the USA and abroad to summarize what is known about SNS and youth and point out gaps in the research base. The research is discussed in terms of education in an increasingly digital world and the challenges and opportunities this poses for educators.

The Challenge of Education in a Digital World

The most popular SNS frequented by youth have been designed with a mostly recreational use in mind. Nonetheless, the technology is relevant to educators. First and foremost, SNS use appears to leap upwards for certain age groups defined around entry into high-school and university (Lenhart, 2007). Secondly, even though the main use for SNS among students is for leisure and socializing, there are trends which suggest the technology may become an important tool in adults' professional lives. This review is meant to give an overview of what is currently known about youth and SNS in order to dispel myths and fears so that educators can begin innovating ways to deploy the technology productively.

Youth and SNS

Although the stages of adolescence and emerging adulthood are periods of intense psychological growth and development, they are distinguished from early childhood development by their lack of linearity and their sensitivity to structuration (Coté, 2009; Kuhn, 2009; Smetana & Villalobos, 2009). While biology determines much of the intellectual growth of young children across cultures, by the time they reach adolescence, pathways of development diverge. In contrast to theories of universal development during adolescence proposed by Piaget, Erikson, and Kohlburg, empirical research has failed to document a single, predictable trajectory for development during this period. Indeed, many adults never achieve certain milestones like full formal operational thinking or complete identity achievement (Coté, 2009; Kuhn, 2009), yet they adapt well nonetheless to the vicissitudes of adulthood. The hardwired programming of development appears to run its course before adolescence and thereafter the human organism faces developmental challenges determined by the institutions of its cultural setting. We are all programmed to learn a language and grasp objects as children, but more elaborate developmental challenges may vary or simply not exist depending on one's cultural milieu in later life (Coté, 2009). For example, the extended developmental challenge of identity exploration and experimentation is enabled by a lengthy period sheltered from adult responsibilities. Such a stage does not necessarily occur for children in developing countries or even youth who enter the workforce directly after high-school.

In the case of the industrialized west, age stands as a proxy for certain institutionalized life stages. Certain rights and privileges like voting, drinking, smoking, driving, university attendance, consenting to sex, entering into contractual agreements—in short, all the trappings of adulthood—are subject to age limitations. For many intents and purposes, then, age is more revealing than other measures of developmental maturity, since the former circumscribes which challenges many youth are facing on their developmental trajectory.

This review is divided into age groups under 18 and 18–25. This decision was made post-hoc after reviewing the bibliography. In the reviewed literature, the tendency among researchers is not only to seek samples which fall on one side of this age divide, but also to pose very distinct research questions about each group. Just as an example, the question of sexual predation or stalking mediated by SNS virtually disappears when researching users over 18 years old. Such a change in research direction is motivated by different legal status accorded by age. Pedophilia is a risk children literally outgrow since the definition is based upon the victim being under a certain age. Older users can experience sexual solicitation online as well, but it is not *ipso facto* a crime based upon the ages of the parties concerned. Other differences in the research agenda trained on these two age groups are elaborated as part of the discussion. These age groupings should be useful to educators since they also correspond to educational life stages: roughly, high-school and university, from which nearly all of the samples were drawn.

The Present Uses of SNS

There are several reasons why educators may take an interest in SNS-use by youth. Firstly, SNS adoption is highest among the youth population (Lenhart & Madden, 2007a). Yet even though youth have been quick to embrace these sites, they do not necessarily have all the skills they need to navigate them safely. SNS can become so complex that they are more like sophisticated software suites as opposed to just a network of Web-pages. One of the main differences is that these suites do not have how-to manuals associated with them and hence, all users are learning to navigate them by trial and error. The subject matter is difficult enough that some instruction may be required and indeed welcomed by students (Wing, 2005).

Secondly, given how sophisticated many SNS are, many users are repurposing the features to do more than simply socialize. SNS are quickly becoming direct competitors to search-engine technology for the dissemination of information. Peers circulate information produced by third parties within SNS, but SNS is also increasingly the site of the creation of original content which is not archived by Webcrawlers. Comprehensive information-seeking on certain topics and personalities may eventually require logging into an SNS. Many politicians, for instance, maintain an active presence with SNS accounts.

The migration of political activity online has been tied to the rise of SNS. From campaign organizers' perspective, SNS grants them instant access to more information on their prospective supporters than ever before. From a supporter's perspective, the peer-to-peer affordances allow discussion, organization, and mobilization within a virtual polis (Silverstone, 2007). Outside the political mainstream, many activists and causes rally around SNS as well. For civics educators, SNS may be a very useful tool to teach.

The potential to engage via SNS has implications for media and arts education as well. One attractive feature of SNS for youth is that their profiles become a place-holder for content they produce anywhere else on the Web (Lenhart, 2007). Such opportunities used to be limited to those with the technical knowledge to host and publish their own Web-pages, but SNS is now being used for the same purposes at much lower cost. As a quick means of claiming an interested audience, SNS may be an incentive to produce multimedia texts outside the classroom.

Closely related to this last point is the identity-work many students do on SNS. In addition to being a portal to their creative output, many youth put as much work into their SNS profiles as statements in of themselves. The opportunity to compose a multimodal text is not always on offer in standard language arts curriculum, yet many youth are doing it in their spare time when they limn their digital identity statements with carefully chosen quotes, music, video, wallpaper, fonts, biographies, and photographs. These profiles also allow for co-construction of identity by nesting a profile within a network of peers (Mallan & Giardina, 2009), something even personal Web-pages did not offer in such an accessible format.

The Future Uses of SNS

Whether the same SNS providers exist in 10 or 20 years, and regardless of whether they resemble their current forms, their impact will still be discernable. SNS have been at the forefront of a bid to align cyberspace more closely with real geographies of people and places. Although there is a tendency to present a somewhat idealized version of oneself online, the face youth are willing to put online is more and more recognizably anchored in their real selves, networked to their real-life friends and nested in a virtual group which corresponds to an offline institution (school, town, corporation). SNS foreground issues of managing one's own reputation online, something which will be germane for everyone in the workforce in the near future. Conducting background checks on candidates' SNS profiles is now routine for employers, as is managing contact with professional colleagues on SNS. Similarly, certain corporations are experimenting with SNS-like features to increase efficiency of communication among their employees. Along with desktop publishing and email, familiarity with SNS may soon be a relevant job skill.

The Current Study

The aim of this review is to gather evidence-based research into the subject of SNS engagement particularly as it affects youth (divided into users under 18 years old and users between 18 and 25). Only research which reported age as a variable was included. Articles were retrieved by performing searches in psychology and communications online databases for the names of popular SNS (and sites with SNS-like features) such as *Facebook, MySpace, LiveJournal, YouTube, Orkut, Bebo, Friendster*, and *Xanga* and then narrowing the returned results to studies on the particular age groups of interest. Evidence of many different forms was accepted, including content analysis of Web-pages and news reports, quantitative surveys, and qualitative interviews. Case studies with only one subject were excluded. Bibliographies from the first wave of results were also used to track non-indexed, published research. In total 76 reports met criteria for inclusion in the analysis. Research is further sub-divided by country of origin, so as to avoid erroneously collapsing different populations together and because the research agendas and priorities appeared to differ between nations.

BACKGROUND

The Rise of Social Networking Sites

SNSs began as a relatively obscure phenomenon online but have since exploded into the mainstream (Boyd & Ellison, 2008). Although the term is nowadays synonymous with the most popular brands (*Facebook* and *MySpace*), SNS encompasses many competing online services, as well as other sites which increasingly partake of SNS-like features. Many rose (and fell) before *Facebook* was even in its development stages (Boyd, 2007) (see Figure 2.1).

Indeed, part of the problem with research on SNS is that the technology itself is a moving target; no two services are exactly the same and even the same service might not be comparable to earlier iterations of itself. Expert Danah Boyd has proposed a general definition of SNS's unique properties which establish some common ground between the different providers:

> We define social network sites as web-based services that allow individuals to (1) construct a public or semi-public profile within a bounded system, (2) articulate a list of other users with whom they share a connection, and (3) view and traverse their list of connections and those made by others within the system. (Boyd, 2007)

In addition to these somewhat novel features, SNS also combine certain affordances common to many other so-called networked publics, including persistence (a record of actions remains), searchability (actions are easy to retrieve), replicability (actions can be copied with no degradation of content), and invisible audiences (actions can be read by unintended viewers) (Boyd, 2007). In many ways, SNS have grafted a socially-based front-end onto existing platforms like email, blogging, chat, and photo-hosting services.

As with most new technologies (Wartella & Jennings, 2000; Wartella & Reeves, 1985), the most enthusiastic users are youth, which has lead to charged debates over both the meaning and impact of these new online environments as well as drafting of policy (Marwick, 2008).

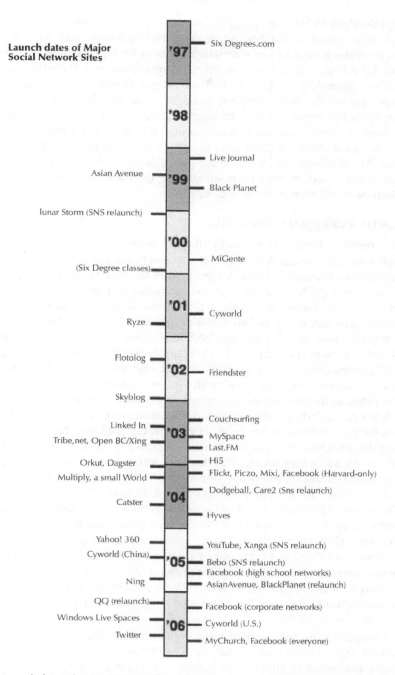

**Launch dates of Major
Social Network Sites**

'97 — Six Degrees.com

'98

Live Journal

Asian Avenue — '99

Black Planet

lunar Storm (SNS relaunch) —

'00

(Six Degree classes) — MiGente

'01 — Cyworld

Ryze —

Flotolog —

'02 — Friendster

Skyblog —

Couchsurfing

Linked In — '03 — MySpace

Tribe,net, Open BC/Xing — Last.FM

Orkut, Dagster — Hi5

Multiply, a small World — Flickr, Piczo, Mixi, Facebook (Harvard-only)

Dodgeball, Care2 (Sns relaunch)

Catster — '04

Hyves

Yahoo! 360 — YouTube, Xanga (SNS relaunch)

Cyworld (China) — '05 — Bebo (SNS relaunch)

Facebook (high school networks)

Ning — AsianAvenue, BlackPlanet (relaunch)

QQ (relaunch) — Facebook (corporate networks)

Windows Live Spaces — '06 — Cyworld (U.S.)

Twitter — MyChurch, Facebook (everyone)

Figure 2.1. Launch dates of major SNS reproduced from Boyd, D. M., & Ellison, N. B. (2007). Social network sites: Definition, history, and scholarship. Journal of Computer-Mediated Communication, 13(1), article 11. Retrieved from http://jcmc.indiana.edu/vol13/issue1/boyd.ellison.

Digitalization in Schools

Concurrent with the rise of SNS as a mainstream technology, many western industrialized countries have achieved near saturation rates for Internet access in schools. The basis for this concerted effort has been the importance of information and communications technology (ICT) skills (defined many different ways) for students both in the present and the future (Mueller, Wood, & Wiloughby, 2008). While this push has materialized in ample access to hardware and connectivity across the board , efforts focusing on support necessary to fully exploit ICT as part of a learning environment have been less thorough (Mueller et al., 2008). If SNS is incorporated into the classroom, it is usually as a threat to be mitigated by safety workshops. Since SNS has been stereotyped as a distraction from schoolwork, there has been less focus on ways to repurpose its affordances to enhance education.

NORTH AMERICAN PERSPECTIVE

The American Perspective: Youth 18 and Under

English-language research on SNS is dominated by American work. Freely available, nationally representative research reports from the Pew Internet and American Life Project give a quick survey of how SNS and other new media are incorporated into teenagers' lives. Teenaged SNS users, for instance, tend to be highly engaged communicators daily across any and all media available to them (cell phones, blogs, instant messaging, email, and face-to-face). SNS also appears to be eroding the popularity of IM and chatrooms (Lenhart, 2007). There are, however, preliminary findings that IM is still preferred over SNS by upper-class youth who have long swaths of uninterrupted computer access to engage "live" with their peers (Zhao, 2009). Teens who create content and distribute it online also engage across more communication platforms, and SNS is a particularly attractive portal to share their work. The Pew study also contradicts the popular concern that online engagement is displacing face-to-face meetings, since the most intense social media users are active in extra-curricular activities and see their friends in person very often (Lenhart, 2007). A complementary report on SNS in particular revealed that membership among teens soars at the juncture when entering high school, from 41% to 61% (Lenhart & Madden, 2007b). The two providers of choice are *MySpace* (85% of users) and *Facebook* (7% of users). A separate study conducted on *MySpace* found that the teenaged cohort also had the largest networks of contacts (25% counting over 91 contacts) but also that a third of *MySpace* profiles are inactive (defined as having only 1 friend or a single log-on) (Thelwall, 2008b). Parents reported whether their teenager had a profile on such a site with 73% accuracy (Lenhart & Madden, 2007b). For the most part, parents are acutely aware of SNS and the supposed risks it entails, but they are far more concerned about the risks than their children (Rosen, 2006).

Comparing the self-reported rates of information disclosure from (Lenhart & Madden, 2007b) and a content analysis of 1475 American teenagers' public *MySpace* profiles performed by Hinduja and Patchin (2008) some striking differences emerge. The much higher rates at which teens post identifiable information about themselves according to the Pew study is likely explained by the fact that the sample is a mix of teens who use private and public profile settings. The site-crawlers used to collect the data for (Hinduja & Patchin, 2008) and (Thelwall, 2008b) could not access private profiles for analysis, 27–39% of the random sample. Hence, such samples of publicly viewable profiles may be more cautious with information. Furthermore, the site-crawler collection of data is insensitive to the intentionality and truthfulness of the information posted. The Pew study found that 56% of teens falsify some of the information on their profiles (Lenhart & Madden, 2007b). Given that the most popular uses of SNS by youth are staying in touch with friends they see often (91%) and friends they see rarely (82%), that is, friends they know in person, the inclusion of false information could be a smoke-screen deployed to throw off strangers. One shortcoming of the Pew survey, however, is that it was unable to compare the profiles of respondents to verify that the users successfully hid their profiles (as opposed to believing that they had).

A final study of American youth by Williams and Merten (2008) focused strictly on the content of teenagers' blogs published publicly to an unnamed SNS. Their sample included one male and one female profile from each of the 50 states. Since this study focused on bloggers exclusively, the coding of profiles had much more raw data to draw from and rates of disclosure were higher. Based on this study, it seems bloggers as a sub-group of SNS users are more likely to reveal personal information by dint of writing so extensively about their lives. This may entail a risk to privacy since the information is publicly available and extremely candid, including the highest rates of revealing the profile owner's full name, potentially embarrassing photos, and phone number (see Table 2.1).

Table 2.1. Sample study of bloggers' profile on SNS.

Content on SNS profile	(Lenhart & Madden, 2007b)	(Hinduja & Patchin, 2008)	(Williams & Merten, 2008)	(Thelwall, 2008a)
Full name	14%	9%	43%	–
Photo (inappropriate photo)	79%	57% (5%)	100% (17%)	–
City	61%	82%	–	–
School	49%	28%	–	–
E-mail address	29%	1%	–	–
IM alias	40%	4%	–	–
Phone number	2%	0.3%	10%	–
Chemical use (alcohol, tobacco, cannabis)	–	2–18%	27–81%	–
Swearing	–	20–33%	44%	38–47%
Privacy settings on	66%	39%	–	–

Although SNS site affordances tend to operate on a privacy binary (where information is either totally exposed or concealed) young users find creative ways to exercise granular controls over their profiles by using existing features in novel ways. A qualitative study by Lange (2008) focusing on teenaged *YouTube* users exposes how youth "seek ways to carve out privacy in highly visible media environments" (Lange, 2008) and are mostly successful.

The subtext of much of the research done on teenagers on SNS has been the issue of unwanted contact from strangers or child "predators". The concern is not unwarranted; according to the Pew study 43% of teen SNS users have been contacted by complete strangers. More than half of teens delete the message or ignore the user, while 20% follow-up by trying to get more information about the person's identity (Lenhart & Madden, 2007b). The evidence of children coming to harm due to this contact, however, is shakier, despite American policy being drafted in response to the apparent threat. Marwick (2008) has analyzed popular press coverage of the panic surrounding youth and SNS and concludes that it has all the same features of previous technopanics:

> Technopanics have the following characteristics. First, they focus on new media forms, which currently take the form of computer-mediated technologies. Second, technopanics generally pathologize young people's use of this media, like hacking, file-sharing, or playing violent videogames. Third, this cultural anxiety manifests itself in an attempt to modify or regulate young people's behavior, either by controlling young people or the creators or producers of media products. (Marwick, 2008)

In her reasoning, the Deleting Online Predators Act (DOPA) currently under consideration by US government is a direct reaction to a mostly imaginary threat posed to youth by SNS, and is proceeding along the exact same lines as the panic over exposure to pornography which culminated in the Children's Internet Protection Act (CIPA).

Marwick's (2008) critique is supported by national research done by Wolak, Finkelhor, Mitchell, and Ybarra (2008) and Ybarra and Mitchell (2008). In interviews with 1,500 youth and over 400 police investigators, Wolak and her colleagues (2008) found that only a small percentage of sex crimes against children were initiated online and none of the crimes involved a forced abduction or stalking a victim based on information of their home/school address. Rather than the popular myth of a naïve child being tricked by a predator, almost all cases were classified as statutory rape with little outright deception on the part of the adult perpetrator. In the authors' opinions, "social networking sites like *MySpace* do not appear to have increased the risk of victimization by online molesters" (Wolak et al., 2008) who by and large still seek a very specific type of susceptible child and expend great effort grooming their victims instead of abducting visible, random online targets. In fact, among the 15% of 10–15 year-olds who had received an unwanted sexual solicitation, SNS's had a lower incidence (4%) than chatrooms (32%) or IM (43%). More importantly, among children unwanted contact online tends to come in the form of online cruelty rather than solicitation (Ybarra & Mitchell, 2008).

Compared to the abundance of research on the dangers of SNS for teenagers (threats to privacy and safety), there is comparatively little which takes an ambivalent or positive stance. Ethnographic studies by Danah Boyd are possibly the best-known example of research which observes typical teenaged engagement with SNS. Teens negotiate with and rework the affordances of such sites to suit their needs. However, these sites also have a degree of power to restructure teens' relationships as well. In the case of adding contacts to one's profile, SNS has provoked an awkward binary choice when articulating one's network of friends on *MySpace* (Boyd, 2006). The ability to nominate eight top friends for public display from among one's network of contacts has presented even more diplomatic challenges to teenagers. However, in both cases teens have been creative and resilient faced by these clunky features, first by downplaying the significance of adding contacts to one's profile and later inventing various new uses for the top friends list to preserve their friendships (Boyd, 2006).

Teenagers have similarly reworked the roles for authorship of their digital identities on SNS, such that they are no longer the work of a single writer. By including friends alongside them in their profile pictures, giving friends with better HTML coding skills the password to edit their profile on their behalf, and cultivating as many public comments from their contacts as possible, teenagers apply a collaborative, wiki style process to building an online identity (Mallan & Giardina, 2009). They gladly cede exclusive control over their online identity to situate it within a group framework instead.

An emerging body of exploratory research is examining what the educational potential of SNS may be among a sample of low-income teenagers. Interviewing 11 diverse teenagers, Greenhow and Robelia (2009a; 2009b) have identified instrumental and emotional support roles played by contacts on *MySpace*. What's more, the site encourages communication via multi-modal texts, which the authors argue is increasingly the norm outside of the school environment but which the formal curriculum has mostly ignored (Greenhow & Robelia, 2009b). Combined with evidence that SNS is not a popular avenue for dangerous strangers to target children, work by Greenhow (2009b) on the educational value of SNS suggests that bans in schools may shortchange learners in the long run.

Marwick forecasts that the storm of fear surrounding teenagers and SNS will subside, but that may be a double-edged sword, as the furor has indirectly driven much of the extant research on teenagers (Marwick, 2008). It remains to be seen if research which takes a neutral or positive stance on SNS and teenagers will outlive the obsession with stranger danger and loss of privacy.

The American Perspective: Emerging Adults 18 and Over

There are many quantitative studies of American emergent adults who use SNS, especially *Facebook*. This is no doubt due to the popularity of using undergraduate samples in survey-based research and the ubiquity of *Facebook* use among American undergraduates. In fact, the universal appeal of *Facebook* to undergraduates is virtually taken for granted. One of the primordial questions in the domain of SNS is what the traits and motivations of non-users are, as well as what motivates users to join different competing services.

Overall, the quintessential non-user appears to be a male (Acquisti & Gross, 2006; Hargittai, 2007; Tufekci, 2008) senior (Acquisti & Gross, 2006; Raacke & Bonds-Raacke, 2008) or graduate student (Acquisti & Gross, 2006) who is less inclined to use expressive/social applications online (as opposed to instrumental/functional ones) (Tufekci, 2008) and is more likely to live off-campus (Hargittai, 2007). The study by Tufekci (2008) used qualitative interviewing and surveys to examine the very specific appeal of *Facebook* as a means for social grooming (Donath, 2008), that is, low-cost signals to sustain relationships and found that this was distinctly unappealing to the typical non-user. However, compared to his peers who do use SNS, the typical users concerns for privacy (Acquisti & Gross, 2006), actual grades (Pasek, More, & Hargittai, 2009), level of shyness (Tufekci, 2008), number of friends (Tufekci, 2008), and computer self-efficacy (Schrock, 2008) do not appear to differ.

Some research has gotten more specific about what motivates users to choose different brands of SNS. Hargittai's (2007) study of a highly diverse sample of undergraduate students revealed that there are reliable differences of race and parental education in which sites users adopt. The composition of *Facebook* users is overall younger, less Hispanic, lives away from home, and has more highly educated parents.

Separate studies on *Facebook* use also indicate that maintaining relationships are an important motivation among its user-base (Pempek, Yermolayeva, & Calvert, 2009; Raacke & Bonds-Raacke, 2008; Sheldon, 2008a, 2008b). *MySpace* users are comparatively more female, more Hispanic, less Asian, and are unlikely to have parents with advanced degrees (Hargittai, 2007). Hargittai concludes that considering SNS use in an aggregated form elides important differences in their user-bases, since in her own dataset when all SNS users are pooled all effects except for gender disappear (Hargittai, 2007).

A content analysis by Jones, Millermaier, Goya-Martinez, and Schuler (2008) of 1400 publicly available *MySpace* pages found the majority of users 18–35 to be white, followed by Latino and then black users. Eighty percent of users self-identified as heterosexual, and 50% single.

Another preliminary study suggests that white *Facebook* users tend to have more homogenous, white networks of friends and as a consequence experience less misunderstanding and higher satisfaction with life (Seder & Oishi, 2009). The authors caution, however, that "In many respects, our results present more questions than answers" (Seder & Oishi, 2009), although it is consistent with research that *Facebook* users are more likely to be white in general (Hargittai, 2007).

Among users of *Facebook, MySpace*, and both services, *Facebook* is rated as a more trustworthy brand (Fogel & Nehmad, 2009) which corroborates findings from privacy research conducted on *Facebook* users (Acquisti & Gross, 2006; Gross, Acquisti, & Heinz III, 2005). As with research on teenagers, a popular theme within the research on emergent adults is the issue of privacy and self-disclosure. However, the focus in research shifts decidedly away from the threat of contact by dangerous strangers and tends toward problems such as identity theft.

One of the earliest studies of a *Facebook* population reported that university students regularly gave away enough accurate information about themselves in their profiles (real name 89%, birthday 84%, address 24%) to put themselves at risk for various kinds of fraud (Gross et al., 2005). Paradoxically, students from the same university were concerned about privacy in general, and wished to avoid theoretical breaches of privacy, yet these fears did not dissuade undergraduates from joining *Facebook* nor posting information they would not want strangers to know about them.

In general, students expressed a belief that other users on *Facebook* were putting themselves in danger by being too cavalier with their privacy (Acquisti & Gross, 2006). These early studies found that students almost never used the privacy settings on *Facebook*; only 1.2% concealed their profile from public perusal (Gross et al., 2005) and 25% were not aware of the privacy features available (Acquisti & Gross, 2006). Based on more recent research which shows about a third of university students now employ privacy settings (Fogel & Nehmad, 2009; Lewis, Kaufman, & Christakis, 2008), it appears that users are becoming more proactive about safeguarding their personal information. An innovative study by Lewis et al. (2008) found a network effect for use of privacy settings in a sample combining *Facebook* and residency data: one of the few predictors of using privacy settings was having a room-mate who used them, too (see Table 2.2).

Table 2.2. Study based on use of privacy settings on SNS.

Content on SNS profile	*(Acquisti & Gross, 2006)*	*(Jones et al., 2008)*	*(Fogel & Nehmad, 2009)*
Full name	89%	11–29%[a]	81%
Photo	90%	75%	86%
Address	24%	–	9%
E-mail address	–	–	35%
IM alias	–	2.7–40%[a]	47%
Phone number (cell phone)	10% (39%)	–	9%

[a]Higher ranges refer to information posted to blog within SNS profile.

Hand in hand with concerns over protecting theftable and reidentifiable personal data for emergent adults have come concerns over managing online impressions of oneself to different social spheres. This same issue plays out for adolescents as well (Boyd, 2006, 2007, 2008), but for older youth the stakes are arguably higher, especially when considering relationships with university professors or the transition into the workforce.

A controlled study showed that students had more positive learning expectations toward a teacher who disclosed more personal information on a mock-up *Facebook* profile (Mazer, Murphy, & Simonds, 2007). A follow-up study showed self-disclosure on part of teachers increased their credibility as well (Mazer, Murphy, & Simonds, 2008). In contrast, preliminary evidence shows that 33% of undergraduate students would not wish their teachers to view their online profiles (Hewitt & Forte, 2006), while 20% would not want employers to see them either (Peluchette & Karl, 2008). In both cases, young women objected more to such "interlopers" on *Facebook*. Such concerns throw into relief the difficulties of showing a single face online to an invisible audience.

What has elicited interest in SNS as a special case within CMC is the fact that users limn a digital body into existence using dense, elaborate presentations of themselves and a host of different strategies. Such presentations of a digital body online used to be limited to Web-page owners but have since exploded into a mainstream activity with low barriers (Boyd & Ellison, 2008). Most importantly, these self-presentations fuse both the individual and their network to form an impression which is not exclusively self-authored. As Zhao, Grasmuck, and Martin (2008) put it: "The appeal is as much to the likeability of my crowd, the desirability of my boyfriend or the magic of my music as it is to the personal qualities of the *Facebook* users themselves." Their content analysis of 83 *Facebook* profiles revealed that co-authored content such as photos of oneself uploaded by contacts and banter posted to a profile's public wall were far more popular than the exclusively self-authored "about me fields" (Zhao et al., 2008). Not only are co-authored statements of one's identity more popular, they appear to carry more weight than self-authored statements depending on the context. Remarks about promiscuity and excessive drinking from one's network increases males' attractiveness but damages females' more than identical, self-authored statements (Walther, Van Der Heide, Kim, Westerman, & Tong, 2008). Statements about a profile owner's attractiveness posted by their contacts also impacted ratings of the owner's attractiveness more strongly. In contrast, self-authored statements about being extraverted were weighed more heavily than contact-authored ones, suggesting that either certain judgments rely less upon outside testimony or that multiple factors are involved (Walther, Van Der Heide, Hamel, & Shulman, 2009).

Additionally, such digital bodies are anchored in one's offline identity more than ever before, and are mostly resistant to outright fabrication of a fictional self by dint of their incorporation in a network which rewards truthful disclosure (Acquisti & Gross, 2006; Grasmuck, Martin, & Zhao, 2009; Gross et al., 2005; Manago, Graham, Greenfield, & Salimkhan, 2008; Zhao et al., 2008). For example, even though *Facebook* includes no field to enter one's race, minority students very vocally claim their cultural identities on their profiles by including inspirational quotes from cultural icons and lists of their tastes (Grasmuck et al., 2009).

Though grounded in reality, online bodies are slightly more elastic performances of identity, and viewers of them tolerate a certain amount of air-brushing. The well-known secret of touching up one's online identity on *MySpace* allows young adults to use it as a safe space to experiment with "hoped-for" selves and get instant feedback

(Manago et al., 2008). One male interviewee described it in the following, forthright terms:

> I could say for certain that there is no one out there with a profile that one hundred percent matches who they really are. If you look nice in a photo then you throw it up there. If you're doing something cool like skydiving, you throw that up there. At least if it's not things you're doing now it could be things that you aspire to do in the future, or if you're making yourself appear as a player or someone poetic, it may not be exactly who you are now but it could be someone who you aspire to be in the future (Manago et al., 2008)

Similarly, an analysis of self-reported profile photo habits on *Facebook* shows that young adults and women are far more likely to update their photo more often, especially if they find a more attractive or recent one (Strano, 2008). The profile photo is thus not archival but a living text which is constantly being revised to make the best possible impression.

The desire to be popular on SNS has been compared to the same desire offline by Zywica and Danowski (2008). In their study, they divided undergraduates into four quadrants based on self-esteem and extraversion. The most socially popular quadrants, both high on extraversion, paid much more attention to their online identity by posting more photos, more flattering photos, removing their name from unattractive photos, and choosing attractive profile photos. Although few participants admitted they made efforts to be popular on *Facebook*, two thirds identified someone else who did. Moreover, both popular and unpopular subjects agreed that those who wish to appear popular add as many contacts as possible to their profiles. It thus appears that for young adults SNS may continue to be an important site for working through identity even after adolescence (Greenhow & Robelia, 2009a; Livingstone, 2008; Mallan & Giardina, 2009; Thelwall, 2008a).

SNS are potentially an excellent source of network data, since linkages are often displayed as part of a user's profile. In general, studies which rely on an individual's network size are more common than studies which do network-level analyses. There is massive fluctuation in the size of contact networks, but the range appears to fall between 135 (Gross et al., 2005; Lewis et al., 2008) and around 200 contacts (Acar, 2008; Ellison, Steinfield, & Lampe, 2007; Fogel & Nehmad, 2009; Lampe, Ellison, & Steinfield, 2007) on average, although all studies which attempt to capture this statistic report many outliers on both ends of the distribution.

Studies that analyze SNS network size have found that extroverted emerging adults (Acar, 2008) as well as those who post their former high-school (Lampe et al., 2007) have larger networks while socially anxious students have smaller networks (Sheldon, 2008b). Larger networks make profile owners appear more physically and socially attractive (Kleck, Reese, Behnken, & Sundar, 2007; Tong, Van Der Heide, Langwell, & Walther, 2008) but this effect plateaus after a point for social attractiveness (Tong et al., 2008). Evaluations of trait narcissism via *Facebook* profiles are also mediated by network size (Buffardi & Campbell, 2008). One study indicates black, Latino, and Indian students have larger networks of contacts (Grasmuck et al., 2009). The authors attribute the size of minority students' networks to the uniquely supportive environment of the university where the study was conducted, and posit that offline clubs with a cultural identity focus translate into more online contacts. Even though the sample may be unrepresentative, the authors' interpretation of the results are insightful because they incorporate the effect of offline networks on the formation of online ones. They are particularly attentive to the blending of online and offline networks; in their opinion, the greater online network sizes of their minority students is a side effect of the many clubs and organizations on campus which bolster cultural communities. In support of this theory, other researchers have found that the networks students maintain on different platforms (SNS, IM, and face-to-face) have substantial overlap, and SNS networks are still biased toward contacts the user has met in person (Lampe, Ellison, & Steinfield, 2006; Subrahmanyam, Reich, Waechter, & Espinoza, 2008).

Similar to the adolescent research, the early waves of SNS use in emergent adulthood found few positive outcomes associated with it. One line of research has treated SNS as a means for enhancing the amount of social capital an individual can maintain (Ellison et al., 2007; Steinfield, Ellison, & Lampe, 2008; Valenzuela, Park, & Kee, 2009) and found promising results, including longitudinal data (Steinfield et al., 2008). In a cross-lagged correlation, they found that more intense *Facebook* usage correlated a year later with a significant increase in social capital, a relationship much stronger than its inverse, suggesting that SNS may be instrumental in the future success of students. One subject summed it up saying:

> I think [*Facebook*] is very good for networking ... it's very good. My high school is very into networking ... I guarantee every single person in the high school will make an effort to maintain those *Facebook* friendships and so that when we're all in our forties, go back to our reunion, and we'll still be able to get in touch with each person we know. You know, 'so and so is a doctor.' And, we wouldn't hesitate to call on them for a favor, just because we went to the same high school ... .(Steinfield et al., 2008)

SNSs like *Linkedin* and *Doostang* are built around the very principal that a network of familiar contacts are better purveyors of information and resources than a massive impersonal archive of information. The success of SNS at tapping and organizing these kinds of benefits, however, has not been extensively researched.

Although there is a greater abundance of research on SNS users aged 18–25, it is of a mixed quality. An issue with the controlled studies is the extent to which they can be claimed as reflective of the "natural" use of SNS. For instance, a study by Wang (Wang, Moon, Kwon, Evans, & Stefanone, 2009) demonstrated a reliable effect of gender and attractiveness, where subjects (especially men) were consistently more interested in meeting attractive opposite sex partners based on a mock-up profile. The study is somewhat problematic because the stimulus controlled for the effect of photo attractiveness by keeping almost all the other fields in the mock-up profile blank. Especially given the findings from (Lampe et al., 2006) that contacts on SNS tend to move from offline-to-online, experiments which use mock-up profiles of strangers (Kleck et al., 2007; Mazer et al., 2007, 2008; Tong et al., 2008; Walther et al., 2008; Walther et al., 2009; Wang et al., 2009) may be quite dissimilar from the mainstream use of SNS which limits the applicability of findings.

Research into emergent adulthood and SNS picks up and continues many of the same themes from adolescent research. Privacy is perhaps the most deeply researched area, although the risks attributed to lack of privacy are very different for each age group. Both age groups have incorporated SNS as a site to do identity work, and both take advantage of SNS affordances to enmesh their presentation of self with their on-line network in many forms. Lastly, considerations of positive outcomes for SNS use are still in a nascent stage for both groups.

THE CANADIAN PERSPECTIVE

So far, Canadian scientists have not published much material on SNS. In the vein of privacy, Christofides , Muise, and Desmarais, (2009) examined self-disclosure on SNS. Although willingness to disclose personal information in real life was the greatest predictor of willingness to do so online, the willingness to disclose online had a non-significant relationship with the desire to control one's personal information. They hypothesize that, at least among students, the risks of retaining tight control over one's personal information may be evaluated as greater than the risks of disclosing (Christofides et al., 2009). This study echoes the findings of Acquisti (Acquisti & Gross, 2006) and Gross (Gross et al., 2005).

The same team of Canadian researchers has also investigated the impact of SNS on romantic relationships (Muise, Christofides, & Desmarais, 2009). On the basis that a partner's *Facebook* profile makes comments and connections visible among a wide circle of friends in a very ambiguous style, the authors correctly hypothesized that more *Facebook* use aggravated feelings of jealousy. *Facebook* was a reliable (if small) contributor to partner jealousy (Muise et al., 2009). This research can be viewed as a specialized application of danah boyd's theories of SNS, namely collapsed contexts and persistence (Boyd, 2007), specifically romantic partners getting mixed into the same sphere as other acquaintances (including exes).

A second team of researchers in Canada has focused on which traits correlate with intensity of *Facebook* use overall (Orr et al., 2009; Ross et al., 2009). When considering very broad personality traits from the NEO-FFI, Ross and his colleagues (2009) report that none of these are as important as computer-oriented motivation and competence when modeling intensity of *Facebook* use, a contradiction to research done by Schrock (2008). However, stable personality traits do correspond to differences in *Facebook* usage patterns. Shyness is associated with longer amounts of time spent on *Facebook,* more positive attitudes towards the service, but simultaneously a smaller network of contacts (Orr et al., 2009). Orr and his colleagues suggest that like other forms of CMC, *Facebook* may be more appealing for shy individuals (Orr et al., 2009). This research complements that done by Tufekci (2008) which finds that shy and non-shy users both subscribe to *Facebook,* and Acar (2008) who finds that extraverted individuals tend to have slightly larger networks of contacts.

In general, the Canadian research is weakened by a few points: all four of the above studies focus exclusively on *Facebook,* all were conducted using undergraduate covenience samples, and the samples were furthermore between 75% and 90% female, which undercuts the generalizability of the findings. In light of findings from

Hargittai (2007) which demonstrate clear preferences of certain groups for different SNS services, such findings require replication in more diverse settings before they can be considered universal.

THE INTERNATIONAL PERSPECTIVE

The UK Perspective

Research in the UK has benefited from national support via Ofcom, which is charged with monitoring media habits of UK. citizens and promoting media literacy. Ofcom has done focused research on children's use of SNS (Ofcom, 2008) as part of a report on internet safety. The sample is nationally representative and concludes that 49% of children aged 8–17 have a profile on an SNS, whereas only 37% of parents confirmed this. Similarly, parents were more likely to report that rules were in place specifically *vis-à-vis* SNS but children confirmed this at a lower rate. Children subscribed more and more often to SNS as they got older, however, younger children were more likely to participate in free online game communities which have started to assimilate aspects of SNS. Among specialized SNS services, *Bebo* is the most popular in the UK. (Ofcom, 2008). This study underlines the generational gap between children and parents, who may be intimidated by these new online communities and feel unable to guide their children through them, which is evident in the USA as well (Rosen, Cheever, & Carrier, 2008).

Probing deeper into how teens self-present on SNS, Livingstone has conducted a qualitative study of 16 teens and their use of these sites (Livingstone, 2008). Interviews from her sample reveal that simply opting out of such sites is not an option for many teenagers, who barter in self-disclosure as a basis of forming friendships with their peers. For young users, the style of self-disclosure is more forthright, and these younger teens change the layouts, palettes, and information on their pages like clothing to constantly revise the statement they make. Older teens use indirect means to express themselves and prefer minimal pages which simply display their position within their network of contacts, making their online persona self and network-authored. Indirect, collective strategies of self-presentation via SNS have been researched by some American scientists as well (Mallan & Giardina, 2009; Walther et al., 2008; Zhao et al., 2008).

Another indirect cue in self-presentation online is language. In an innovative cross-cultural comparison of youth *MySpace* profile content, Thelwall (2008) analyzed profiles of both American and UK. subjects to discover that while male teens in the USA use significantly stronger profanity, this gender gap has disappeared in the UK. among young users. Thus, behaviors captured on SNS may be a useful barometer of over-arching cultural trends.

The difficulty of policing SNS content, in the form of swearing or otherwise, in addition to its potential to give strangers access to children has so far kept the technology out of schools in the UK. Turning to teachers' role as guide through various Web 2.0 platforms, the BECTA research group reports findings from a sample of 27 schools and 2600 students. Among this population 75% reported having an SNS profile, and reported that their computer use, especially for Web 2.0 activities like SNS, took place outside school (Crook & Harrison, 2008).

In contrast to parents' hesitance to engage with SNS, some teachers are interested in harnessing student interest in these social channels to teach, but meet resistance from administrators who restrict school internet access out of concern for child safety (Crook & Harrison, 2008; Sharples, Graber, Harrison, & Logan, 2009). Among teachers and administrators the consensus is that an approach which keeps SNS (and other Web 2.0 technologies) accessible from school while teachers empower children to navigate them successfully is ideal, but not yet feasible.

Although SNS offer the potential for unknown adults to contact children, another creative UK. study shows that this potential is going largely unmet. By comparing the contact networks of 50 teenagers and 50 elderly adults, Pfeil (Pfeil, Arjan, & Zaphiris, 2009) found that even though teenaged users have significantly larger pools of contacts on *MySpace*, the ages of their contacts are almost totally homogenous and close to their own. In other words, teenaged *MySpacers* associate almost exclusively with peers of the exact same age. Nonetheless, the potential for contact from dangerous adults still weighs heavily on school administrators. This same emphasis on risk and harm is prevalent in US educational policy (Marwick, 2008; Ybarra & Mitchell, 2008).

In contrast to the research on young learners, studies of university students' *Facebook* use in the UK. hold out the possibility of positive uses for SNS in a higher education environment. In a survey of 213 undergraduates, 50% registered a *Facebook* account at their university before the first semester began in order to get a head-start making friends at school. Madge (Madge, Meek, Wellens, & Hooley, 2009) pose that many register a *Facebook* account at this juncture because of the twofold benefit of previewing classmates at one's new school and staying in touch with friends from one's hometown. Students were so positive about these benefits that the authors suggest administrations could encourage *Facebook* pre-registration to smooth the transition into university life. Nevertheless, they caution school staff to keep a certain distance from *Facebook* (Madge et al., 2009), which students still prefer to claim as a semi-private space (West, Lewis, & Currie, 2009). Indeed, although a minority used *Facebook* for strictly educational/instrumental purposes (Madge et al., 2009), a content analysis of wallposts between all students in an undergraduate faculty conducted by Selwyn (2009) shows the importance of SNS as a "backstage" to university life. While students avoid using *Facebook* to contact staff or teachers, and they rarely use it as a site of outright learning, it serves an important function as a commons where they can collectively work through the vicissitudes of student life and seek support.

While American researchers have treated the issue of teachers and staff appearing on *Facebook* (Hewitt & Forte, 2006; Mazer et al., 2007, 2008) we were unable to find American research comparable to the above which considers the positive educational impact of SNS for university students. Although *Facebook* networks began grounded in offline, brick, and mortar universities, most American research does not consider the possibility of an SNS contributing to educational goals in its corresponding real-world space.

Overall, the quality of research on teenaged and younger users of SNS in the UK. is bolstered by national interest from Ofcom, and strikes a balanced perspective between managing the risks and opportunities posed by SNS. Research done on older youth is confined to undergraduate *Facebook* users, and although it combines different methods (qualitative interviews, surveys, content analysis) the body of research is not large enough to draw sweeping conclusions about the experience of SNS for youth aged 18–25 in the UK. Both the studies from Lewis (Lewis & West, 2009) and West (West et al., 2009) are based on samples of 16 students each. Although the Selwyn (2009) study observed a sample of 909 users, it was 75% female and 215 profiles were closed off from public viewing. The emergent literature, however, approaches SNS from angles not considered by the American body of research and opens paths to alternative avenues of inquiry.

Non-Anglophone Perspectives

English-language research on SNS and youth from non-Anglophone countries is rare. This is understandable since most SNS are geographically bound and research on those networks is most pertinent within those borders. A Dutch study using a sample of 881 10–19 year-olds subscribed to *CU2* (a popular Dutch SNS) found that use of the site could enhance well-being and self-esteem (Valkenburg, Peter, & Schouten, 2006). In general, positive comments left on users' profiles correlated with better self-esteem (0.48) which, in turn, correlated with increased well-being (0.78). The authors hypothesize that since SNS users pick and choose who is allowed to post comments to their profile, the tendency is to select only one's friends who are more likely to leave compliments. Interestingly, neither the overall size of any child's online network nor the frequency of feedback was related to the increase in self-esteem, merely the positive valence of comments. American research often includes self-esteem as a predictor variable in research on SNS (Acar, 2008; Dong, Urista, & Gundrum, 2008; Ellison et al., 2007; Steinfield et al., 2008; Zywica & Danowski, 2008), but rarely investigates whether SNS can enhance self-esteem or well-being.

A qualitative study of SNS use in Japan yielded interesting findings *vis-à-vis* the cultural boundedness of different SNS providers. In interviews with Tokyo college students, Takahashi (2008) found that half her subjects were users of both the Japanese *Mixi* service and *MySpace*, although they used each one differently. It was generally agreed that *Mixi* was meant for one's uchi, or closest, most exclusive circle of friends, and that the content of a *Mixi* profile was written less for the owner's personal expression but to affirm bonds within that group. Unfortunately, the technology's capacity to maintain larger uchis than possible in the past, combined with the collision of contexts, placed massive strain on profile owners to blog simultaneously for multiple audiences. By turn, multi-account users felt less restrained on their *MySpace* profiles, since they were on a foreign SNS which encourages an individualistic style of self-presentation. Rejecting a real-life friend request on *Mixi* was unthinkable, but on *MySpace* it was perfectly acceptable. American scholar Danah Boyd has researched the costs and benefits of contact requests in an SNS setting (Boyd, 2006) but very few studies have gathered samples with multi-account users (Fogel & Nehmad, 2009; Hargittai, 2007) and none have looked at their motivations for using multiple services.

Interpretation

The earliest foundations of SNS research for youth age groups have focused on patterns of uptake. Although there are still some youth who abstain, uptake of SNS is so popular the question is no longer if they use SNS but rather which one. Even Internet users who do not sign on for any of the well-known SNS will probably see similar features appear more and more frequently elsewhere online. As a platform, SNS appears to have taken hold for the majority of young Internet users.

What distinguishes SNS from other forms of CMC is the willingness of users to disclose personal, identifiable information about themselves. This has driven research on both adolescents and emerging adults. Although the hazards change to reflect the life-stage of users, both veins of research have a negative valence especially in the USA, where SNS are treated as breaches to personal security which are liable to end in stalking of adolescents or identity theft for emergent adults.

At the same time, research has also shown that users tend to be aware of the risks yet are willing to assume them. This can only be understood when the potential rewards of self-disclosure are counter-balanced against the risks. SNS users are not necessarily being duped out of their private details for nothing. To some extent, the social lubricant of SNS requires self-disclosure to gain benefits. Research which emphasizes positive aspects of SNS-use redeems members who sometimes get painted as gullible victims.

Some of these rewards have been underlined for emergent adults such as amassing more social capital and support, especially at difficult junctures like transitioning between high-school and university. The UK body of research has elaborated this latter aspect the most for emergent adults. However, as far as adolescents are concerned, the benefit of maintaining friendships, a central developmental task during adolescence, has gotten short shrift as a worthwhile reason to barter in one's personal information. The Dutch study by Valkenburg and her colleagues was the only study which allowed room for a positive outcome for adolescents via SNS (Valkenburg et al., 2006). In the case of adolescents and children, fears for their personal safety (no matter how far-fetched) appear to trump all other arguments. As with any social science endeavor, there is a heavy normative tone to the existing body of research on youth and SNS which treats the phenomenon first and foremost as a problem in need of management and intervention.

Research which studies SNS engagement by youth for its own sake from an ethnographic approach hints at some of the constructive identity work adolescents and emergent adults do on these sites. This vein of research has not yet been translated into quantitative, large-scale research which makes it difficult to compare against risk-based research. Indeed, the very nature of identity-work may not ever be easily reduced to a quantitative analysis without compromising the constructs under investigation.

What can be drawn from ethnographic research is the fact that, although SNS creators envision a certain use for their sites, users are resilient when faced with seemingly coercive requests for information. Responses like deploying smoke-screen answers to required fields, deleting or ignoring messages from strangers, and using multiple accounts to manage multiple identities, are just some of the ways users retain granular control over their online persona beyond the sometimes inadequate privacy settings offered by sites. The fact that adolescents have molded SNS more to their liking is an example of the versatility of the medium, which educators should consider when attempting to mobilize ICT in school settings.

Summary of Major Differences

Overall, American research has focused on questions of who uses SNS and why, including demographic differences. Canadian research has treated some of the same issues. The risks of SNS-use under investigation change to reflect different life stages between age groups. Quantitative research has focused on potential risks and hazards based on disclosure of personal information, whereas American ethnographic research has taken a more neutral stance and examined practices online. There is comparatively little work which a priori entertains the possibility of benefits from SNS use, but much of it comes from abroad (the UK, the Netherlands). So far, Japan has produced the only study which thoroughly examined the subjects' spectrum of identities across multiple sites, despite evidence from the USA that many users maintain accounts on more than one SNS.

DISCUSSION

Recommendations for Educators/Administrators

As professionals who work with young people on a daily basis, educators need to at least experiment with SNS environments to gain some familiarity with them. At the very least, such hands-on experience may quell anxieties about what takes place on such sites. Ideally, it can inspire ways of repurposing the technology for educational goals. Compared with other websites which host static information and deliver it top-down to an anonymous, invisible audience, SNS unlocks the potential for peer-to-peer communication and feedback. More importantly, the commercial instances of these sites appear to have hit upon a mix of features and designs that appeal to users and encourage copious, spontaneous participation. There is much that could be emulated from these sites when designing virtual learning environments.

While online bulletin board systems have similar discussion features, they do not foreground each user's digital body as strikingly as SNS, which may be the key to the appeal of such sites for adolescents who invest so much effort in creating an online/offline identity. SNS is particularly supportive of peer-to-peer dialogue that is highly visible, persistent, and organized around the axes of identities. Just as an example, such environments would be conducive to soliciting peer feedback on drafts of written work. SNS are also a low-cost means of including guests and experts from outside class on an ongoing basis. A central feature of SNS popular with youth is the performance of taste in music, fashion, and movies; the same affordances could be reworked to include novels and plays. In all the above cases, SNS add another layer of peer-to-peer learning beyond the top-down, teacher-centered model. The appeal of using the SNS framework is that portfolios of student work, comments, and tastes are organized as part of their digital bodies which contributes to the larger project of building their personal identities.

However, one problem of the existing commercial platforms is that they are an ambiguous blend of public, private, and commercialized space. Youth are already wary of interlopers such as parents, teachers, and future employers trespassing on what they consider their territory. Furthermore, for youth who abstain from commercial SNS either for personal reasons or parental rules, absorbing commercial SNS for class activities may be controversial. There are additionally legal issues of making students register for certain sites which collect and sell personal information on youth which could be a legal conflict of interest depending on the school's position toward commercialization. Although commercial SNS are a good template of features which work well for peer-to-peer dialogue and feedback, the question of ownership of these sites remain problematic. Therefore, schools who wish to adopt SNS may consider open-source or other alternatives which they can own and moderate.

Technical and ethical issues aside, another concern is adapting teaching practice to take advantage of SNS affordances. Previous efforts to bring ICT into the classroom have fallen short of expectations mainly because the arrival of technology has preceded planning on how it will be incorporated usefully into an already heavy courseload. Without this kind of clear direction, ICT including SNS can go unused even after schools have invested in the technology.

For educators who wish to focus on interventions for students' safe, recreational use of SNS, the evidence on how online solicitation occurs must displace imaginary worst-case scenarios of violent abduction by child predators. Research has demonstrated that youth are not unilaterally at risk for grooming by child predators and, more importantly, the youth who do become involved with online stalkers rarely perceive the relationship as exploitative or inappropriate. Safety interventions should reflect the proportional incidence of each threat, such that predation falls behind more common dangers like leaking embarrassing information or being misled by SNS-based advertising.

FUTURE RESEARCH

Future research could benefit from starting in a neutral or ambivalent stance toward SNS. This would allow both advantages and drawbacks of the technology to emerge from results. Additionally, SNS studies are frequently focused very specifically on certain SNS brands, as if SNS use were a uniform construct. Such research may not be useful in the long-term if the different aspects of SNS are not exploded during analysis, especially as new sites take up and discard different features (anonymity, articulating ties, reciprocity, digital bodies, wikidentity, circulating information) in new configurations. Research which clings too closely to specific SNS without attention to the individual affordances risks becoming obsolete should the associated site go out of business or even revamp itself, as often happens.

A related issue is the focus on user-profile dyads. Although many commercial SNS would prefer their users to be exclusive to one brand, the reality appears to be that many users manage more than one profile, sometimes across different sites. Site-crawling studies or studies which foreclose probing into the full range of a user's online identities are analyzing a truncated view of the individual's online presentation.

In the case of site-crawling studies, the data is even harder to interpret because the intentionality or truthfulness of a profile cannot be assessed without recourse to the author(s).

CONCLUSION

In the context of education, SNS are sometimes styled as an intolerable distraction from learning, if not an outright threat to student safety. Although bans on such sites have the appearance of a proactive response from administrators, this may short-change students in the long run. In the case of search engines, a slightly older technology, recent research has shown that undergraduates who grew up with the technology use only the most basic features and use poor heuristics to assess the credibility of results (Hargittai, 2010; Hargittai, Fullerton, Menchen-Travino, & Thomas, 2010). The belief that such digital literacy will take care of itself informally for youth is a myth which is liable to play out again with SNS. Youth enthusiasm for SNS must not be mistaken for competence.

Commercial sites like *MySpace* and *Facebook* may dominate the public imagination when speaking of SNS, but they are merely two examples of the genre. SNS can be repurposed for many goals which serve educational goals.

CONCLUDING REMARKS

Like *Friendster* before it, the future of *Facebook, MySpace*, or any SNS is uncertain. However, even if individual brands fall out of favor or go out of business altogether, the general schema of SNS structures and affordances will retain their appeal. Indeed, many other sites have hastened to absorb SNS-like features into their architecture because they have proven so popular. Social networking is on the cusp of becoming a banal, everyday technology, which is a huge achievement for such a new form and makes it a pertinent object to study.

KEYWORDS

- **American Perspective**
- **Facebook**
- **Social networking sites**
- **Web-pages**
- **YouTube**

Chapter 3

Perspectives on Learning Design

M. Mahdavi and A. K. Haghi

INTRODUCTION

The complex process of teaching and learning requires complex, multi-faceted models of implementation. One tool will not meet all needs in all contexts. Changes impact and influence existing models—rendering yesterday's solutions obsolete. In the field of learning, an adaptive model of technology selection and governance is required to ensure that all stakeholders' needs are met. A solution today may not be accurate tomorrow. A sustained process needs to be enacted to align context changes with changes and approaches to learning methods and technologies available. The university must adapt, using technologies and models of understanding, in this case to reconcile teaching, research, IT, a changing environment, financial accountability, and managerial models. Learning management systems (LMSs) have a position in higher education (certain types of under-graduate level learning are more structured and focused on memorization or content exploration). To meet the needs of all learners in various stages of their education, a multi-faceted (holistic) view of learning must be considered. Increasingly, personal learning environments (PLEs) provide the tools and model to attend to the diverse learning needs of individuals today. An LMS/LCMS handles and controls the delivery of self-paced, instructor led, and e-learning courses. The LMS lets you publish courses and place them in an easy to use online catalog. Learners log into the LMS with any browser, select courses from the course catalog and take them. The LMS tracks the learners' activities, providerships, professional licenses, progress, compliance with course assignments, and much more. The LMS provides online reports for each course and learner yet is inexpensive and provides the essential features you need.

LMSs enable effective design and delivery of learning materials. They enable tools for authors (instructors) to design learning materials that might include multimedia. They also enable learner activity management in the learning process. Moreover, they provide tools for effective and efficient assessment of the learners. In this chapter we study learning activity management systems and the main components that such a system should provide in order for instructors and learners to effectively participate in the learning process.

There is a growing trend among medical educators towards the use of new learner-centered teaching and preparation methods, based on self-learning and with specific objectives. In contrast to conventional theory lectures, they are more efficient in promoting learning, are more flexible both for teacher and pupil, and moreover, they help the learner to acquire the self-learning habit, which should become a daily practice over the course of the learner's professional life. In relation to computer aid and

Web-based learning, there is extensive literature that shows how the computer is effective in the instruction of engineering professionals in comparison with conventional education, especially programs which include problem-solving or interactive methods. In engineering self-education, it facilitates the learner's attention, allows individualized progress and provides immediate non-competitive and flexible feedback, adapted to individual needs. Engineering is the discipline of implementing of science and mathematics to develop explanation for the problems, and to find practical solutions that have an applicable outcome. Engineering is the knowledge of creating a new world based on the rules which science has already discovered. LMSs are now regularly considered mission-critical to the long-term success of most organizations, extending beyond the learning domain to include the convergence of learning, performance management, staffing, certification, providing just-in-time education to customers, providing instant access to learning content through 24/7 learning portals, and ultimately measuring the effects of all types of learning to help meet organizational goals and objectives.

With all that said, the fundamental question still exists: "So, what is the best LMS?" And, which one should we use in our organization? It is a question that we have continually been asked since we began evaluating LMS solutions. In our earlier reports, we made attempts to "short list" some of the best LMS products based on the maturity and range of functionality offered in each system. While these lists did assist somewhat in showing which systems were ahead of the LMS evolutionary curve, the lists themselves offered only limited help for the purpose of choosing a system that most closely matches an organization's specific critical needs and business requirements.

The purpose of this online LMS service is to provide an answer to this question for you. Just do not expect to see a pre-defined list of "The Best LMS Solutions" here. The selection and benchmarking process really does hinge on what you are trying to accomplish: your specific needs; your organization's limitations (e.g., infrastructure requirements, requirements for back-office integration, etc.); to some degree, your own subjectivity; and even more esoteric issues such as cost versus functionality. We understand that you face a multitude of budgetary and business requirement issues.

The information is also valuable for those who are studying the dynamics of the LMS market. In addition to extensive research data on LMS core and advanced functionality, the LMS research database also includes additional information about the types of users that choose each system, the typical size of end-user companies, pricing averages for different scenarios, which vertical markets are most likely to use the LMS, and so forth.

In recent years, computers are widely used for learning purposes. They enable using multimedia for effective delivery and also the Internet for accessing the learning materials. Electronic learning (E-learning) refers to the use of computers for delivering learning materials. LMSs provide a set of learning materials and enables users connect to the system, normally through the Internet. The LMS is a Web-based software application or used to plan, implement, and assess a specific learning process. Typically, the LMS provides an instructor with a way to create and deliver content, monitor learners' participation, and assess their performance. The LMS may also provide learners with the ability to use interactive features such as threaded discussions, video conferencing, and discussion forums. Learning Activity Management Systems (LAMS) are flexible learning design tools that enable instructors organize and monitor learning activities of the learners. These activities include assignments, quizzes, and also collaboration. Collaborative learning is a learning process during which learners collaborate with each other for solving problems. In other words it promotes team work in the learning process. Assessment is also an integral part of such systems that enable instructors effectively evaluate learners' activities in the learning process. This chapter explores LMSs and their key components that enable instructors organize and monitor learning activities of the learners. It also explores challenges and issues that exist in this field.

Many classroom activities are sequential and structured, and teachers are used to this kind of learning design. However, online learning environments are often disorderly and uncontrolled, and offered to students as a free smorgasbord. The chapter describes a bridge between the two learning environments through the use of a Web-based LAMS. In the cases described in the chapter structured, flexible Web-based tasks are provided which can be used both for online self-study and in a classroom setting. The study has found preliminary evidence of success in gaining the acceptance of students and teachers for the ideas of adding control to online learning activities while adding flexibility to classroom teaching. Obstacles and challenges have been recorded in the cases and are converted into practical ideas for improvement.

BACKGROUND: LEARNING ACTIVITY MANAGEMENT SYSTEMS (LAMS)

The Learning Activity Management System (LAMS) is a learning design system with a particular focus on sequencing of collaborative learning activities. This chapter reviews a number of lessons learned from the development of LAMS, and their implications for both the existing IMS learning design specification, and its future scope and purpose. It proposes a number of areas for further development.

Current collaborative tools in LAMS include: question and answer (with student answers shared with the group either anonymously or identified), polling (with total responses shared with the group), asynchronous discussion forum, synchronous chat, notice board (simple text content/instructions), resource presentation and sharing (URL/Webpages/files), notebook/journal, assessment submission, MCQ and True/False (with options to display feedback, average class score and "high" score), and various combinations of tools, including "chat and scribe." In addition, a grouping tool allows any tool to run in either whole class or small group modes. The flexibility of learning design indicates that many more collaborative tools can be developed to broaden the impact of the LAMS platform. Sequences created in LAMS can be shared among teachers either via e-mail or the LAMS repository. LAMS is a fully functioning system currently undergoing beta trials with a range of K-12 school and university partners across Canada.

Constructive learning is a process in which the learners construct new thoughts, ideas, and concepts making use of their knowledge and experience (Beatty, 2003). It provides greater opportunities for the learners to have the responsibility and control over the learnt material. However, there is a key issue of education in the learning process which is face-to-face interaction with instructors, classmates, and the coordinators rather than simply interacting with content (Alexander, 2008). In order to have a constructive learning, switching back to traditional education it is not suggested though. In fact, the face-to-face benefits of the traditional systems should be incorporated in the e-learning systems. LAMSs appeared to satisfy the demand. They combine benefits of e-learning system with controls of traditional educational systems which result in an effective learning system. However, some aspects has to be further investigated.

Nor Azan (2007) has developed an LAMS system named SPAP. Their system works based on a concept called Conceptual Model of Level A IMS LD where IMS LD is a specification used to provide a containment framework that can describe any design of a teaching and learning process in a formal way. SPAP is an example of a LAMS that allows teachers to plan, manage, and monitor learning activities. It also enables learners to carry out learning activities planned by teachers. Activity tools developed in SPAP are based on teaching methods: discussion, project, problem-solving, group teaching, and simulations. There are seven activity tools in SPAP authoring module: set induction, grouping, discussion, searching, assignment, project, and comment.

SPAP includes five modules: registration, synopsis, authoring, learner, and monitoring. The registration module is used for registering various users, organizations, courses, and users to courses. This module can be accessed by an administrator. The synopsis module is for setting course objectives, duration, topics, teaching methods, and learning outcomes for different topics. The authoring module lets the teachers design learning activities in a sequence and save the design. Figure 3.1 shows authoring interface of SPAP. For customization, activities are dragged from the left frame into the middle frame and the links between activities are created using the given tools. The learner module enables learners to perform learning activities that are designed by the teachers. Learners are given a clear visual path of their progress via the progress bar which is visible at all times. Figure 3.2 shows the learner module. The monitoring module enables monitoring the students' progress in performing learning activities at any time. SPAP provides a new learning experience for learners while helping teachers to manage learning activities. In general it gives learners and teachers a positive perception of security, suitability, accurateness, and satisfaction.

Dalziel (2003) explains a simple example of sequences based on a definite question. It was initially designed for history students around the ages 14–16 in a K-12 school context. It was designed for a group size of around 20–30 students, potentially located in more than one physical location. The sequence lasts for 4 weeks:

> Week1: Students enter an asynchronous discussion environment, and discuss and debate the conceptual questions. This activity encourages students to articulate their views, and to engage directly with their peers.

> Week2: Students are given access to content objects (e.g., text documents, Webpages, files, etc.) and URL links to relevant Internet Websites. After reviewing the contents, students use a search engine (e.g., Google) to find an example of a Website about the defined topic. Students then share this URL (and a comment about why

they selected it) with the class, so that all students are able to view each other's selected Websites.

Week3: Students are randomly divided into multiple small groups and each group is given a live chat environment to discuss some specific questions about greatness from the teacher. One of the students is assigned the role of "scribe," and is given a special scribe interface where they can record the small group's discussion on the specific questions. Then agreements of each group on a definite topic are sent to the class Webpage, where all students can see the outcomes from each of the small groups.

Week4: Students individually write up a report on the original question, based on their learning experience across the whole sequence. This report is submitted to the system, which helps managing the workflow of marking and commenting for the teacher. The end of the sequence is reached when students receive their marks and comments from the teacher.

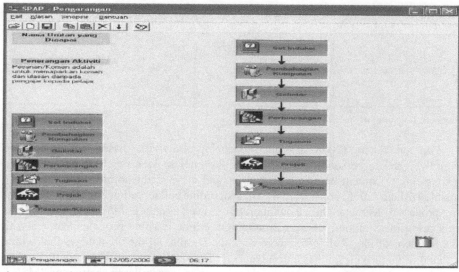

Figure 3.1. Authoring module of SPAP.

One of the powerful features of the learning activity approach is that the content of a sequence can be easily changed to suit a different discipline, while leaving the activity structure unchanged. The learning activity sequence can provide a "pedagogical template" which may be useful in many contexts by changing the "content" to suit different discipline areas. The focus on easy re-use with LAMS means that these changes can be implemented and ready to run with a new student group. Moreover, the pedagogical template itself should be modifiable, so that if a teacher wishes to change the order of the tasks, or add/subtract activities from the template, this can be easily achieved. In LAMS, this is possible within the authoring environment using a simple drag-and-drop system which helps make explicit the teaching and learning processes as a series of discrete activities.

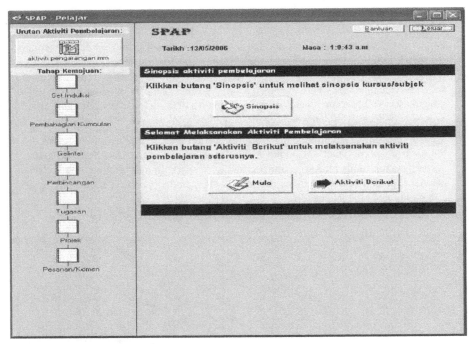

Figure 3.2. Learner module of SPAP.

Fan and Jiaheng (2008) describe sequencing of activities based on ontology and activity graph in personalized learning environments. As a result, the creation and sequencing of content-based, single learner, and self-paced learning objects would be enabled. Mann (2008) considered the issue of online collaboration impact on the learning process of learners. Online collaboration is used as a part of their learning process in sharing ideas among learners. In addition, online learning provides learners with the freedom of physical and time constraints. Despite all the advantages that online collaboration provides, it has a disadvantage. That is, learners involving in the online collaboration process may find themselves in a situation similar to chatting on "MSN," which is more in line with socializing instead of learning. In other words, learners may misuse the online environment so that the learning might not take place. As a result, the purpose of online learning may be eliminated. Therefore, the environment should be less flexible. In other words, monitoring tools may be useful in order to lead learners to learn when they perform a learning activity. LAMS environment also provide a live monitoring tool for educators in order to monitor the progress of students so that inappropriate behaviors of misusing the collaboration tool in a destructive way can be minimized.

Alexander (2008) described a thorough examination of a comprehensive LAMS. According to their claims, 3200 members in 80 countries use this system. Using LAMS, digital course plans are shared and adopted and the experiences are exchanged easily. This open-source software manages the e-learning activities in three modules:

teachers, monitors, and learners. This system can work as a stand-alone e-learning system or in combination with other Virtual |Learning Environment (VLE)/LMS such as Course Maker (http://www.cs.nott.ac.uk/~smx/pgche/coursemarker.html). Course Maker is a Web-based package for creating course content. It provides a visual environment for creating and changing the course design. It also provides a number of pre-designed course templates that can be used for creating interactive multimedia learning materials such as image maps and quizzes. Other examples of such systems include Sakai (http://sakaiproject.org/index.php?option=com_content&task=view&id=103&Itemid=208), LRN (http://dotlrn.org), Moodle (http://en.wikipedia.org/wiki/Moodle), BlackBoard (http://en.wikipedia.org/wiki/Blackboard_Inc), and WebCT (http://en.wikipedia.org/wiki/WebCT).

As mentioned before assessment is an integral part in learning systems. Web work (http://webwork.csis.pace.edu/webwork) is an Internet-based system for generating and delivering homework problems to students. Its goal is to make homework more effective and efficient. Each Web work problem set is individualized. That is, each student has a different version of each problem, for example the numerical values in the formulas may be slightly different. Students complete the assignments, log onto the Internet and enter their answers using a Web browser. Web work system responds student by telling them whether an answer (or a set of answers) is correct or not and also records whether the student answered the question correctly or not. The student is free to try a problem as many times as he wishes until the due date.

A key educational benefit of this system is that if a student gets a wrong answer, the student gets immediate feedback while the problem is still fresh in his mind. The student can then correct a careless mistake, review the relevant material before approaching the problem again, or seek help (frequently via e-mail) from friends, the teaching assistant (TA) or the instructor. It increases the effectiveness of traditional homework as a learning tool by:

- Providing students with immediate feedback on the validity of their answers and giving students the opportunity to correct mistakes while they are still thinking about the problem.
- Providing students with individualized versions of problems, meaning that instructors can encourage students to work together, while still requiring that each student develop an answer to his own version of the problem.

It also increases the efficiency of traditional homework by:

- Providing automatic grading of assignments.
- Providing information on the performance of individual students and the course (or section or recitation) as a whole.

LEARNING ACTIVITIES AND ROLES IN LAMS

In this part we discuss the properties of LAMS and its components. One of LAMS benefits is that staff uptake is widely affected by the level of institutional information and communication technology (ICT) provision. Higher levels of pupil motivation are expected using the coherent, integrated, and structured LAMS compared to traditional courses. Moreover, the self-paced LAMS environment encourages students with anonymous favors developing their confidence, autonomous learning, and meta-cognitive skills. As a result, the users of such systems become more inclusive from the traditional ones. The possible activities in a LAMS are not limited, but we can name a number of standard and common tasks and activities in current e-learning systems. Some of the most important activities are:

1. *Informative Activities*—These activities include those that play an informative role for users. More important activities include:
   - *Notice board:* It is a straightforward way of providing content to the learners (as a bulletin of the professors). It can be implemented in two methods. Firstly, a brief link based legend index of items can be presented. This index points to different targets. Secondly, a larger area can provide as much content as possible. It might also use links. The content can be rich text, html, links, word documents, audio, video, and so forth.
   - *Share resources:* The learners can share their resources using this activity.
   - *Task list:* This activity allows learners to inform the monitors and the authors to complete the tasks. It may be an individual activity to manage the activities too.

2. *Evaluative Activities*—These activities include those that are used in the evaluation process. Some of the important activities in this category include:
   - *Multiple choices:* This activity is a quiz consisting of several questions each of which containing several items to choose from. The learners should select the correct answer of each question in the specified time. The multiple choices are designed for each part of the course according to the needs of that part.
   - *Submit files:* The learner's work or answers to exercises can be received by the system in different ways. One of these ways is by submitting files. Learners can make files from their work and submit them in the specified area for assessment.
   - *Research activity report:* The learners should inform the instructors and the research activities/results of their current tasks.

3. *Collaborative Activities*—There are a number of activities that enable collaboration between learners. Some of these activities include:
   - *Chat:* It is a popular activity for the learners. They can chat together, to the monitor/instructor based on their needs.
   - *Scribe:* It is the reports that the learners write.
   - *Forum:* It is an asynchronous discussion environment for the learners. The discussion threads usually are created by the authors.
   - *Group communications:* The learners joining together as a group can communicate with each others. For example, file and resource sharing between them can be done.

4. *Reflective Activities*—These activities reflect the learners ideas or thoughts during the learning process:

- *Question and Answer (Q and A):* The learners can ask questions and the instructors answer them. The important aspect is that all questions and answers are presented in a single page visible to all.
- *Notebook:* It is an activity for recording learners' thoughts during a sequence.
- *Survey:* It collects the learners' surveys and opinions about a topic.
- *Voting:* It allows the learners to vote on a specific idea in the sequence, topic, or learning object.
- *Learner bulletin:* It is a place for the learners to inform the others about anything in the learning space.

LAMS KEY FEATURES AND FUNCTIONS FOR DIFFERENT ROLES

In LAMS, there are different roles participating simultaneously such as educational roles, controlling roles, advertising roles and so on. But in all of them, there are three common roles constituting the main LAMS foundation: Learner (Student), Monitor (Assistant), and Author (Instructor). In the followings we describe these roles:

Learner (Student)

The students can study and learn what an instructor wants in a controlled manner. In fact, a LAMS works just like the instructor is asking questions, assigning tasks, teaching, and briefly monitoring the answers to the tasks/assignments to be done. These tasks including reading the provided lessons (in the form of plain text, html, media), chat, scribe, voting, exam, and other previously mentioned activities are not limited to these tasks and can be extended to any other digital contributions in the future.

The overall tasks/assignments can be shown and classified as progress bar or sequence of tasks. Each task in the sequence is represented by a specified name, shape, and color. For example, as in Alexander (2008), the completed tasks are shown with blue circles. The in-progress tasks are indicated with red squares and the tasks that have not been reached are shown with green triangles. However, this is only an assumption and can include more comprehensive states and situations. The task progress bar (or task sequence) navigation can be facilitated with interactive user-friendly navigation tools including exposure to text, tables, images, video, and audio.

In the lower level, after selecting each task, the learner can contribute to the learning process specified by the instructor. For example, if the activity is an html scribe task, the learner should write something specific. If the activity is a multiple choice, then the learner should choose between the multiple choice(s). A notebook tool is also available for the students to write statements to be viewed entirety or collated into a journal of entries available to be viewed by the instructor. An example of different possible areas for the learners is shown in Figure 3.3. Moreover, Figure 3.4 shows a sample activity arrangement.

Progress Bar	Main Panel
The activity icons (for the tasks to do) appears colored and in order here.	*The learning objects in the form of text, pdf, word documents, html, audio, video, etc. are displayed here. The learner interacts with the system in this part.*
Notebook	
The learner notes (for him/her or the author) appears	Start Chat.... Logout

Figure 3.3. An example of different possible learner areas.

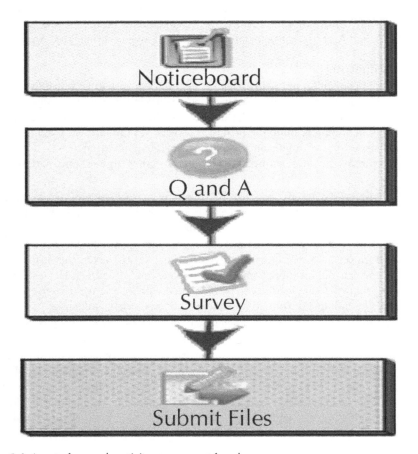

Figure 3.4. A part of a sample activity arrangement for a learner.

Monitor (Assistant)

A key feature in LAMS is monitoring. By this feature, the instructors (or their assistants) can control the students' learning. There are three major parts for this feature: Lesson, Sequence, and Learner.

Lesson-related monitoring activities: In this section, the instructor (or his assistant) can control the overall properties of each lesson and the students that use them. He can also disable lessons, archive them, and so forth. The important properties of this part can include:

- View the lessons' status;
- View the statistics of each class, its learners and their selected lessons;
- View the number of online learners for the lesson at each time;
- View the total number of learner logins for a lesson in a period of time; and
- View the sequences' status of the learners for the lesson.

Sequence-related monitoring activities: A sequence is a chain of tasks or activities for the learners as defined by the instructors (see Figure 3.5). The monitoring tasks for a particular sequence are:

- Grouping the learners (performed by the monitors);
- Assigning activities to the groups;
- Choosing scribes;
- View the overall status of the learners in each sequence;
- View the total number of learners doing each task (online or offline);
- Refreshing the system to be synchronized with the sequences' status of the current learner;
- Controlling the learners such as opening or closing gates, force moving the learners to a specified activity (by moving their icon to the desired activity) or moving out the learners from an activity or a sequence; and
- Making minor sequence content changes by adding or removing particular activities.

Learners-related monitoring activities: Monitoring the learner's progress is an important aspect of LAMS. In the traditional systems, it is constituted from several face-to-face aspects during the learning period. However, in those systems, the details are usually overlooked. The learners-related monitoring tasks include:

- View the overall learner's progress;
- View the learner's sequence progression (completed, current, and new activities that have not been undertaken);
- View the detailed status of the learners in each sequence (according to the selected learner);
- Edit learners' permissions to the sequences (or parts of them) created by the authors (instructors); and
- Export portfolios for a learner or a group of learners in groupware activities.

Some activities such as *Chat* or *Forum* can be revisited provided they have not been instructor-locked and exited with *finish* button. The actual content is displayed on the main panel; in Figure 3.6 the current task is an imported *Wikipedia* text.

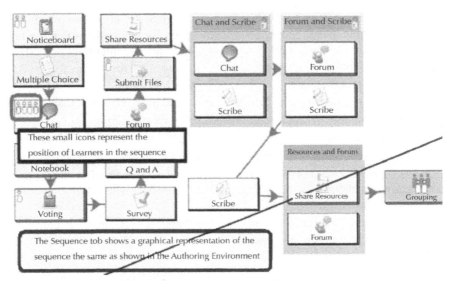

Figure 3.5. A sample of a sequence.

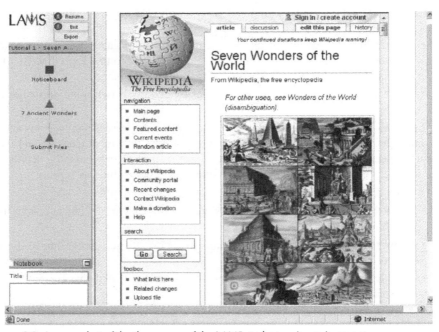

Figure 3.6. A screenshot of the three areas of the LAMS student main environment.

Author (Instructor)

Each activity in a LAMS should be started by or under supervision of the author. In fact, creating multiple fascinating pathways for the learners is not a troublesome task today, but having meaningful consistent set of controlled structures with unified educational goals in such a system makes it hard to implement, considering the collaboration between different parts of the system. In such a system, an author can control the system in these ways:

- Creating, storing, modifying, and reusing the learner activity sequences;
- View a preview of their authored sequences;
- File operations (new, open, save, and save as);
- Editing tools (copy–paste and cut–paste);
- Sequence tools (transition, optional activity, flow, gates, grouping, branching, and preview); and
- Importing and exporting sequences to be shared with colleagues.

A pilot project: Distance education programs are becoming more common in traditional higher education institutions. Some universities are offering courses from their online programs to on-campus students, while others are establishing special programs for their students. LMSs are essential for content development and management of such programs. This case study discusses the Eastern Mediterranean University Learning Management System (EMU-LMS) developed while the EMU-Online program was offered to on-campus students, outlining both technical details, and administrative issues.

In the last 10 years, many universities and higher education institutions have started offering distance education courses for their on-campus students. This serves a number of purposes:

1. The university gains experience on online course development;
2. The university gains experience on the management of online programs and this possibly leads to the establishment of an institute/center specialized in distance education methods;
3. The students taking online courses return quick feedback on the program; and
4. The distance education institutes/centers train a sufficient number of teaching staff/assistants who are qualified in developing distance education programs.

Considering the importance of all these purposes, and believing that distance education will play a more important role in the education system, in the year 2000, the EMU, has started the EMU-Online program, mainly targeted for on-campus students. In the first five semesters, a total of 10 courses were taught, each one a number of times and more than 1000 students have taken these courses. In the current semester six courses are offered and 283 students are registered to these courses.

Considering the requirements and feedback coming from EMU-Online instructors, the EMU Distance Education Institute (EMU-DEI) has developed an LMS called EMU-LMS. The main requirements for developing such a LMS were:

1. The instructors needed an announcement management tool which would list announcements on course work and assignments to students for certain durations;
2. The institute needed to keep track of which students were registered to courses, and for security reasons, assign passwords for access to online course material;
3. The instructors needed online support for assessment, as online quizzes;
4. The instructors needed an online discussion program;
5. The instructors needed a discussion board system to start threaded discussions on course work;
6. The students and the instructors needed support for online submission of course work and assignments; and
7. The instructors wanted to see the "online attendance" of students throughout the semester.

Hence, the LMS seemed to be inevitable to use Internet efficiently and to create a higher level of interaction between the instructors and the students. At this point, there were two options: (1) Purchase the LMS available in the market, or (2) Develop the LMS in-house, with the university's human and financial resources. The decision was to develop an in-house system for the following reasons:

1. To provide better solutions to problems encountered in the EMU-Online program, quickly, and interactively;
2. To develop a low-cost LMS as the funds allocated for the program were limited;
3. To develop a scalable system, this can expand as new requirements were met and as new funds were available;
4. To be independent of private companies which may ask for extra financial resources for upgrades, or which may stop supporting their software in case of financial trouble; and
5. To compete better with other universities using in-house built tools open for further development.

Eventually, with the experience and expertise gained from the EMU-Online program, a vocational online diploma program on Information Management is offered over the Internet in the 2002–2003 academic years for the first time. The program has a one week face-to-face orientation at the beginning of first year. The first group of students who are registered to this program have recently visited EMU for the orientation program.

This section discusses the main properties of EMU-LMS, making comparisons with commercially available LMS products whenever it is appropriate. In the following section, various modules of EMU-LMS will be explained. In Section 3, a comparison of EMU-LMS and prominent commercial LMS products will be provided.

MODULES OF EMU-LMS

In this section, various modules of EMU-LMS will be discussed. For each module, the following information shall be given: (1) The module name, (2) The user-types that can access that module, and (3) A short description of module functionality. Comparisons, whenever appropriate, will be made with two commercially available and widely used LMS's: the WebCT and BlackBoard systems, and with a similar LMS developed at the Middle East Technical University called Net-Class. For some modules, advantages and disadvantages of EMU-LMS compared to other LMS programs will be listed.

User Enrollment

The module handles basic requirements of the LMS such as adding a new user (student or instructor) to the system, modifying properties of existing users, and deleting users. Arranging the user accounts properly increases the system efficiency. This module is used by administrators and instructors (partially). Users can be assigned to individual courses. If a user has no course assigned, he/she stays in the system until the administrator deletes the user to a course, or assigns courses to that user.

In both BlackBoard and WebCT, users can be added to the system over the Internet. In EMU-LMS, this operation can be done by the administrator. Assignment of already existing users to courses can be done by instructors.

Category Management

With this module, courses can be categorized and put into special folders for further references. This helps to group the courses according to the programs in which they are taught. This module is used by administrators. It provides an opportunity to group together courses of a program.

Course Management

The main properties of a course like the number of students, the instructor, or the name of the course can be modified in this module. The Course Management module is linked to some sub-modules. These sub-modules are:

1. Assignment Management, which helps instructors in posting up assignments on a course. Problem/question files can be uploaded to the server, to be downloaded by the students.
2. Quiz Management, with which instructors can design online quizzes for assessment, typically one at the end of each chapter.
3. Announcement Preparation, which is a tool for automatic announcement/deactivation of announcements coming from instructors with deadlines associated.
4. Discussion Board, which is essential for threaded discussions that will be started by instructors. Students respond by posting their opinions/findings on the issue. This module is one of the most important tools for interactivity.

Courses can only be created by the administrator. This makes the system stable and reliable. Instructors can access all the sub-modules.

Lecture Management

Instructors can create lecture notes, course outline, or syllabus by using this module. One advantage of this module is that, instructors can prepare their lecture notes independently, without any need for learning a new editor, using general purpose editors with which they are experienced, such as Macromedia DreamWeaver, Microsoft FrontPage, or Home Site. They upload files prepared with these editors to the site. Broken links in the navigation bars of Webpages are avoided by this method, because all the links are kept in the database. Some uploading tools are implemented in BlackBoard and WebCT, but they are not aimed for lecture management. Lectures are prepared in their built-in HTML editors.

Quiz Management

EMU-LMS supports four types of questions in a quiz: (1) multiple choice, (2) true/false, (3) essay, and (4) fill in the blanks. This tool uses a question database which keeps quiz questions of all courses. An instructor can import questions from this database into a new quiz. The tool also keeps a student list and quiz grades. Individual and overall quiz averages are computed automatically. The instructor can see whether a student has taken the quiz and is graded, has taken the quiz but is waiting for grading, and has not taken the quiz yet.

WebCT has five, BlackBoard has seven different question types. A general question database is used in both systems. Exam duration, answer feedback, and multimedia file attachment support are also included in both systems. BlackBoard has password-protection feature for quizzes.

In EMU-LMS, two types of quiz evaluation are used, student-by-student and question-by-question. This is an advantage for an instructor, because it makes the evaluation easier. Exam time limit check is not implemented in EMU-LMS. Time limit may be an important parameter for an online exam, it helps to measure the students' performance with a certain time limitation.

Announcement Preparation and Handling

Announcements are important guidelines for online students. The instructor can announce exam dates, online chat hours, information about upcoming events, or results of assignments by using this tool. Scheduled events can be posted and removed using provided posting/removal days. When the deadline passes, the announcements disappear automatically. Announcements can be added, modified, or deleted by instructors.

BlackBoard and WebCT uses a time-schedule for announcements. Both systems give the right to post announcements to the instructor. BlackBoard provides an opportunity of automatic announcement creation for events such as quizzes, and grade announcements.

Automatic announcement creation is not implemented in EMU-LMS. On the other hand, allowing the instructor to post announcements can reduce the work of the administrator.

Assignment Management

Assignments have an important role in measuring the performances of online students. There are three type of assignments supported by EMU-LMS: (1) one-to-all—one assignment is given to the whole class, (2) multi-to-all—multiple assignment topics are given to the whole class. The students can choose one of the topics given. One topic can be chosen by more than one student, and (3) one-to-group—assignments are given to groups of students. File uploading is also available, and is optional for all three types of assignments. The instructor can enable or disable file uploading while choosing the assignment type. Group work for assignments is allowed in BlackBoard, but not in WebCT. Tracking of students on assignments is supported in both systems. Assignments can be scheduled for later use.

The major advantage of EMU-LMS is automatic creation of a forum to discuss topics related with the assignment. The moderator of that forum is also automatically assigned while students are taking the assignment.

Discussion Board

The discussion board tool is helpful for the students to start a discussion on a course-related topic, to ask questions, or to share their ideas with the other students taking the course. Administrators have full permission (like creating a forum, deleting messages, etc.) over discussion boards of all courses. The instructor can only control the discussion boards of his/her own courses. The students can have limited control on specific forums if they are moderators of those forums.

In the discussion boards of BlackBoard and WebCT, modification of a specific message is allowed. A search opportunity is provided through the forum. An indication of new postings is helpful for the users. One advantage of EMU-LMS forum management is the automatic creation of forums according to the assignments. When an assignment is given, a forum is automatically created and reserved for discussions on that assignment.

Student Tracking

Two types of student tracking are possible in EMU-LMS. One is tracking the time spent in the pages of a course, and the other is recording the time between a student's login into and logout of the system. The administrator can see student tracking details of all the courses. The instructor can see only his/her courses' student tracking details. User name and surname, how many times he/she accessed the online course, and how much time he/she spent in the course and/or the system are shown to the administrators or the instructors. With this tool, users who are logged on can get a list of current users. Online users are grouped according to the place where they are working.

CONCLUSION

Online education is still in a highly preliminary stage regardless of its extensive acceptance in many fields or disciplines in higher education. It is conceivable that faculty members will attempt to build up from their traditional teaching experiences, especially when there is lack of practical guidance on how to carry out online instruction. As much as one might encourage instructors to be creative and transformative, such innovations take time to develop. People often feel comfortable building upon what they are familiar with instead of starting out creative. While such transitional stage from transfer to transformation is understandable, organizations should look for the best practices to help shorten this transitional process. Such organizational support not only improves the local program quality, but also helps expedite the overall development of distance education by providing more successful teaching and learning cases to the field. This study tries to provide a fairly holistic picture on which instructional technologies and activities are used to promote interactions in the current stage of online education.

In this chapter we studied LAMSs as a tool for designing, managing, and delivering online collaborative learning activities described components and key features of these systems. These systems enable more effective delivery of learning materials in terms of the ratio of learners to instructors. They also enable more efficient assessment of learners' activities as a result of using software tools. Moreover, these systems are time and location. However, there are challenging issues in such systems compared to traditional learning systems where face-to-face conversation, meetings, scientific tours, and so forth are common. As a result, such face-to-face activities might be necessary in parallel to these systems when necessary.

The outlined approach to e-learning necessitates a focus on students, providing them with tools to support their self-governed, problem-based, and collaborative activities. Using a management system, personal tools, and social networks differs from the sole use of an integrated LMS. The approach differs in terms of focusing on empowerment of students as opposed to management of learning. An approach focusing on empowerment of students implies thinking in terms of tools rather than in terms of systems. The idea is first and foremost to provide students with a variety of tools for their self-governed and problem-based activities; to empower students, offering them tools for independent work, reflection, construction, and collaboration. Second, the approach suggests facilitating students' engagement in different networks. Existing social software tools such as weblogs, wikis, and social bookmarking can be used to support e-learning activities. However, these tools are not developed for educational purposes, which mean that a directed effort is necessary to develop educational social software tools to support learning activities.

The perspectives for the use of social software in the form of personal tools and social networks to organize e-learning go beyond a single course and the educational institution in which the students are enrolled. The use of computers to assist learning also enables the formation of social contacts that would otherwise be impossible in learning. Students from widely dispersed groups are able to form online groups. The chapter has argued that personal tools and networks support self-governed, problem-based, and collaborative learning processes. This way of using social software also equips students with valuable tools for using the Web as a resource in order to develop their understanding and solve problems—whether in school, at work, or in their private lives. This has particular relevance in relation to lifelong learning. This scenario for lifelong learning is similar to describes as "learning networks". Although the concept is more organized, it too describes the creation of learner-centered and learner-controlled networks for lifelong learning: Self-organized learning networks provide a base for the establishment of a form of education that goes beyond course and curriculum centric models, and envisions a learner-centered and learner controlled model of lifelong learning. Working as proposed here, students not only learn a specific topic, but also they are equipped with tools to navigate and make active use of the Web to solve future problems. After the end of a course or an education, the networks continue to exist. Continued participation in social networks and creation of new networks give people access to a vast number of people and other resources.

After almost a decade of the LMS experience, educators, and administrators are beginning to question the prominence of an LMS. LMSs are often viewed as being the starting point (or critical component) of any e-learning or blended learning program. This perspective is valid from a management and control standpoint, but antithetical to the way in which most people learn today. LMSs like WebCT, BlackBoard, and Desire2Learn offer their greatest value to the organization by providing a means to sequence content and create a manageable structure for instructors/administration staff. The "management" aspect of the LMS creates another problem: much like we used to measure "bums in seats" for program success, we now see statistics of "students enrolled in our LMS" and "number of page views by students" as an indication of success/progress. The underlying assumption is that if we just expose students to the content, learning will happen. Two broad approaches exist for learning technology implementation:

1. The adoption of a centralized learning management approach. This may include development of a central learning support lab where new courses are developed in a team-based approach—consisting of subject matter expert, graphic designers, instructional designer, and programmers. This model can be effective for creation of new courses and programs receiving large sources of funding. Most likely, however, enterprise-wide adoption (standardizing on a single LMS) requires individual departments and faculty members to move courses online by themselves. Support may be provided for learning how to use the LMS, but moving content online is largely the responsibility of faculty. This model works well for environments where faculty have a high degree of autonomy, though it does cause varying levels of quality in online courses.

2. PLEs are a recent trend addressing the limitations of the LMS. Instead of a centralized model of design and deployment, individual departments select from a collage of tools—each intending to serve a particular function in the learning process. Instead of limited functionality, with highly centralized control and sequential delivery of learning, a PLE provides a more contextually appropriate toolset. The greater adaptability to differing learning approaches and environments afforded by PLEs is offset by the challenge of reduced structure in management and implementation of learning. This can present a significant challenge when organizations value traditional lecture learning models.

NOTE

1. Professors can generate stop points somewhere in a sequence by gates. Gates halt the process of learners through a sequence until a specified condition is met.

KEYWORDS

- **Electronic learning**
- **Learning Activity Management Systems**
- **Learning management systems**
- **Personal learning environments**
- **Web-based learning**

Chapter 4

A Framework for E-Learning in Digital World

B. Noroozi and A. K. Haghi

INTRODUCTION

In recent years we have witnessed an increasingly heightened awareness of the potential benefits of adaptivity in e-learning. This has been mainly driven by the realization that the ideal of individualized learning (i.e., learning tailored to the specific requirements and preferences of the individual) cannot be achieved, especially at a "massive" scale, using traditional approaches. Factors that further contribute in this direction include: the diversity in the "target" population participating in learning activities (intensified by the gradual attainment of life-long learning practices); the diversity in the access media and modalities that one can effectively utilize today in order to access, manipulate, or collaborate on, educational content or learning activities, alongside with a diversity in the context of use of such technologies; the anticipated proliferation of free educational content, which will need to be "harvested" in order to "assemble" learning objects, spaces and activities; and so forth.

This chapter provides information for the revision of engineering curricula, the pedagogical training of engineering faculty and the preparation of engineering students for the academic challenges of higher education in the field. The book chapter provides an in-depth review of a range of critical factors liable to have a significant effect and impact on the sustainability of engineering as a discipline. Issues such as learning and teaching methodologies and the effect of e-development; and the importance of communications are discussed.

The impact that Web-based learning will have on users is powerful. The Web is a communication channel with the ability to use it with audio, video, graphics, and text. Users can then communicate through one of these channels and keep in contact with one another, by groups and even in real time. Training programs benefit greatly from the use of the Web for several reasons. These pros consist of training being able to be distributed quickly and easily, graphically dispersed students can communicate and learn effectively, and the update of materials can be a fraction of the coast of them being revised by other means. Some of the things that attribute successful management of a Web-based training (WBT) include an ability to articulate the value that the Web brings to training development, aligning the program to the corporate culture, and the effect it has on the organization and also understanding the psychological effects of working with an WBT. The ability to work and manage WBT programs is becoming a norm to most companies. The Internet has influenced almost every aspect of business. With the growth in technology and use of the Internet, training professionals and realizing the plethora of uses that can be applied to existing training programs.

Despite the many benefits of WBT programs, there are drawbacks that must be taken into consideration. Some of the disadvantages may be that fact that online activity may be time consuming, implementation and software updates may be very costly, and the psychology behind WBT may disenfranchise people. Because of these factors, sometimes it is better to continue using traditional training techniques or a combination of both. However, it all boils down to the company and asking themselves if this is something that should invest in. Typical questions to ask may be to what is the nature of the performance deficiency or learning opportunity intended to address, who is the target audience and how will they benefit from it and another may be to ask how will it affect the budget of the company. So with these factors in mind companies need to analyze their current situation thoroughly and weigh out the advantages and disadvantages before commitment.

Online learning is not about taking a course and putting it on the desktop. It is about a new blend of resources, interactivity, performance support, and structured learning activities. The key benefits of e-learning solutions are:

- Increased quality and value of learning achieved through greater student access and combination of appropriate supporting content, learner collaboration and interaction, and online support.
- Increased reach and flexibility enabling learners to engage in the learning process anytime, anyplace, and on a just-in-time basis.
- Decreased cost of learning delivery, and reduced travel, subsistence costs, and time away from the job.
- Increased flexibility and ability to respond to evolving business requirements with rapid roll-out of new and organizational-specific learning to a distributed audience. Often an e-learning program will involve development by a number of people simultaneously. Therefore, design and development of standards need to be put in place to ensure consistency and transferability of skills. Students are also far less forgiving in terms of inconsistency of user interface and ease of use than with classroom material.

Designers and developers of online learning materials have variety of software tools at their disposal for creating learning resources. These tools range from presentation software packages to more complex authoring environments. They can be very useful in allowing developers, the opportunity to create learning resources that might otherwise require extensive programming skills. Unfortunately, a number of software tools available from a wide variety of vendors produce instructional materials that do not share a common mechanism for finding and using the resources. A number of organizations started advocating standards for the learning technology. Few thousands of courses are being offered on the Internet. However, most of the courses offered on the Internet are not having good quality. There is need for standards to define the framework for online learning. Standards are desirable for interoperability, convenience, flexibility, and efficiency in the design, delivery, and administration. They provide consistent online dimension for all courses being designed so that all authors/faculty are able to customize the online materials with minimal lead-time. Standards also impose a certain order on chaos resulted due to proprietary products from different vendors. This order provides more uniform and precise access and manipulation to e-learning resources and data. The appropriate standards result in the following benefits:

- Industry-wide standards for learning technology systems architectures.
- Common, interoperable tools used for developing learning systems.
- A rich, searchable library of interoperable, plug-compatible learning content.
- Common methods for locating, accessing, and retrieving learning content.
- Standardized, portable learner histories that can be transferred with the learner over time.

The definition and adoption of complete and sound e-learning standards will help the e-learning market achieve some key goals:

- *Users* of e-learning applications will be able to shift between programs and platforms, to find those that fit their needs with minimal transition cost. Moreover, the learners gain in flexibility since the attained knowledge can easily migrate to future e-learning platforms that follow the same standards. To put it simple, once the user familiarizes with a standardized e-learning technology it becomes easier to familiarize with any variation of this technology.
- *Learning content producers* will focus on the development of content in a standard format instead of developing the same content into many formats for different platforms and applications.
- *Tool vendors* will not spend money for the development of interfaces that glue their tools to e-learning platforms and systems. Lower development costs imply less expensive tools of better quality and subsequently an increase in the size of the potential market.
- *Application and platform designers* are able to choose from a large storehouse of reusable content, systems and tools, and assembled a competitive and effective e-learning platform. They can also populate the storehouse with new modules of content and applications.

The main objective of this work is to define the framework for the development of global e-learning standards that support interoperability of e-learning systems. To achieve this, it is essential to understand the innermost of e-learning process: its lifecycle and its infrastructures as presented in the next section. Next it is necessary to judge on the usability and re-usability of existing work on standards and present the general steps of our approach. Finally, it is important to understand how standards adhere to the interoperability of e-learning systems. This is explained in the fourth section, which presents the most important e-learning interoperability standards. In the conclusion we present a roadmap for the creation of widely accepted e-learning standards. Ernest Boyer (Boyer, 1990) states that, ... *Scholarship means engaging in original research, it also means stepping back from one's investigation, looking for connections, building bridges between theory and practice, and communicating one's knowledge effectively to students.* Therefore, the scholarship of teaching engineering, seeks to find effective ways to communicate knowledge to students. The realization that traditional instructional methods will not be adequate to equip engineering graduates with the knowledge, skills, and attitudes they will need to meet the demands likely to be placed on them in the coming decades, while alternative methods that have been extensively tested offer good prospects of doing so (Rugarcia, Felder, Woods, & Stice, 2000).

Engineering is the profession in which knowledge of the mathematical and natural sciences gained by study, experience, and practices are applied with judgment to develop ways to utilize economically the materials and forces of nature for the benefit of mankind. Engineering is a unique profession since it is inherently connected to providing solutions to some expressed demand of society with heavy emphasis on exploiting scientific knowledge. In the real world, engineers must respond to sudden changes. The engineers of today, and in the decades ahead, also must be able to function in a team environment, often international, and be able to relate their technical expertise to societal needs and impacts. Yet, we start at making transformative changes in our educational system. Our educational challenge is itself a design challenge—making the "right" engineers for our nation's future. The basis for the reform of engineering education is made up of unique experiences, traditions, and everlasting values of specialist training at universities. Engineering educators have to focus on market demand and stop defending the obsolescent and obsolete programs. In order to prepare engineers to meet these new challenges, engineering training and education must be revised and modernized. Today's engineer cannot be merely a technician who is able to design the perfect bridge or the sleek skyscraper.

Today's engineer must not only have a breadth and depth of expertise, but also must be able to communicate effectively, provide creative solutions with vision, and adapt to ever-changing demands. Today's engineer, like any other modern professional, must be someone who can see the big picture

INTEGRATING "WHAT" INTO "WHY" IN ENGINEERING EDUCATION

Web-based education (WBE) is currently a hot research and development area. Benefits of WBE are clear: classroom independence and platform independence. Web courseware installed and supported in one place can be used by thousands of learners all over the world that are equipped with any kind of Internet-connected computer. Thousands of Web-based courses and other educational applications have been made available on the Web within the last five years. The problem is that, most of them are nothing more than a network of static hypertext pages. A challenging research goal is the development of advanced Web-applications that can offer some amount of adaptively and intelligence. These features are important for WBE applications since distance students usually work on their own (often from home). An intelligent and personalized assistance that a teacher or a peer student can provide in a normal classroom situation is not easy to get. In addition, being adaptive is important for Web-based courseware because it has to be used by a much wider variety of students than any "standalone" educational application. A Web courseware that is designed with a particular class of users in mind may not suit other users. Since the early days of the Web, a number of research teams have implemented different kinds of adaptive and intelligent systems for on-site and distance WBE. The goal of this chapter is to provide a brief review of the work performed so far in his area. The review is centered on different adaptive and intelligent *technologies*.

A professional needs to recognize the "why" dimension, as well as the "what" in order to provide a wisdom and understanding. Also, for the profession to attract students there needs to be an enhanced community respect for engineering. This can be assisted if we integrate a person-centered and nature-respecting ethic into engineering education (Hinchcliff, 2000).

The urgent need to change the teaching method of the current engineering education system was the reason for which the author launched to a new plan. The new plan envisaged changes in the curriculum to meet the demands of the industry, now facing strong competition as a consequence of the recent technological changes. With this aim, the authors developed the courses considering following issues:

1. *"Why"* not try replacing one quarter of the lectures with an online resource? As part of online resources, lecture-based courses are taught at many institutions using videotaped lectures, live compressed video, television broadcasts, or radio broadcasts (CSU, 2004; Forks, 2009; UI, 2004). The student can have an easier time communicating online as opposed to in a full classroom (Purcell-Roberston & Purcell, 2000). In addition, student is at the center of his/her information resources. The content information is not delivered as a lecture for the students to hear but rather as information for the students to use. Students are free to explore and learn through their successes and sometimes failures. Instead of the lecture, students could spend time over a month working through some online materials complete with self-tests, interactions, mini-project, or whatever. We can use the replaced lecture time to manage the online course and deal with queries. The presentation would be varied without too much change at once.

2. *"Why"* not replace half the lectures with a simple online resource, and run smaller seminar groups in the time, the lectures would have used (obviously not practical if you lecture to 200 students), but if you online lecture to 40, this might give you a chance to do something different with them face-to-face. When all of a course's lectures, readings, and assignments are placed online, anyone with a computer and Internet access can use online resources in the course from any location, at any time of day or night. One institutional goal of this movement toward computer-aided education based on online resources is to make higher education more economical in the long run through an "economy of scale." If all of lectures, syllabi, and assignments are digitized and put online, tutors could spend less of their time teaching a larger number of students, and fewer of those students would be on-campus using the university's resources (Frances, Pumerantz, & Caplan, 1999). According to the survey results conducted in university of Wisconsin–Madison, the vast majority of the students took advantage of the fact that lectures were online

to view material in ways that are not possible with live lectures. When asked if it was easier to take notes to understand the material when viewing lectures on electronic resources than it would have been attending the same lecture live, almost two thirds (64%) of students agreed, either strongly (27%) or somewhat (37%) (Foertsch, Moses, Strikwerda, & Litzkow, 2002).

Online education of engineering students requires more than "what" the transfer of the knowledge and skills needed by the profession because should always keep in mind that:

- A classical lecture is not the best way to present materials any more.
- We have to demonstrate something visual this is difficult for everyone to see in a lecture.
- We have to work through a simulation or case study which would be better done at a student's own speed.
- We have to ask regular questions during the presentation of content and we have to monitor responses.

FUNDAMENTAL ASPECTS

There are a number of technologies whose integration into modern society has made dramatic changes in social organization. These include the Internet and its attendant promises of more democratic information access and distribution, including distance learning. A totally effective education and training environment, when applied to information technology instructional strategies that are enhanced by the World Wide Web, will include factors that have long been identified as contributing to an optimal and multi-dimensional learning context—a personalized system of instruction (Wild, Griggs, & Downing, 2002). The ingredients of such a system have long been known to contribute to an optimal learning environment for the individual student (Overview of e3L, 2005).

Electronic-based distance learning as a potential lever play an extraordinary role in the creation and distribution of organizational knowledge through the online delivery of information, communication, education, and training (Wild et al., 2002). Numerous studies have been conducted regarding the effectiveness of e-learning. To date, there are only a few that argue that learning in the online environment is not equal to or better than traditional classroom instruction WBT also provides education access to many non-traditional students, but if applied to replace existing educational spaces, the potential effects of this replacement must be evaluated and assessed (Coleman, 1998; Kolmos, 1996; Overview of e3L, 2005). However, e-learning is not meant to replace the classroom setting, but to enhance it, taking advantage of new content and delivery technologies to enable learning. Key characteristics of online learning compared to traditional learning is shown in Table 4.1 (Wild et al., 2002).

Table 4.1. Knowledge presentation consideration for E-learning.

Characteristics of Traditional Learning	Characteristics of Online Learning
Engages learners fully	E-learning should be interactive
Promote the development of cognitive skills	E-learning should provide the means for re-
Use learners previous experience and existing knowledge	pletion and practice
Use problems as the stimulus for learning	E-learning should provide a selection of pre-
Provide learning activities that encourage Co-operation	sentation style
among them members	E-learning content should be relevant and
	practical
	Information shared through E-learning should
	be accurate and appropriate

Interactivity

E-learning should be interactive. There are opportunities for both learners and knowledge holders to build on the information being conveyed in the online environment. Opportunities include threaded discussions, chat areas, and exercises that invite learners to interact with the content and respond. Most importantly, learners and instructors should have a means to contact the content expert or others in the actual learning environment. E-mail, discussion bulletin boards, groupware environments such as Lotus Notes (TN), and so forth, can be used to support interactivity.

Repetition and Practice

E-learning should include a means for repetition and practice. In other words, the courses should engage and challenge the learners to evaluate, select, and use the information in their everyday lives. The content should be relevant to the learner's frame of reference (i.e., content that is practical and understandable to the user). Case studies, simulations, and "what would you do" exercises help learners grasp the content and find ways to use the new information creatively in their lives. The Web is an excellent resource for establishing a frame of reference for exercises.

Presentation styles: E-learning should provide a selection of presentation styles. The most beneficial courses offer several ways for learners to absorb the material. Written content is fine, but more learners grasp concepts with illustrations that accompany the content. Video, audio, and other multi-media choices within an online environment contribute to the richness of presentation choices. Instructors can inquire in the online classroom about other ways to present the material if alternative delivery methods are available or preferred by members of the learning group.

Content: E-learning content should be relevant and practical. Adults learn better when the objectives of the course are directly linked to issues, theories, case studies, research, and knowledge that is practical. Simply putting material online does not make e-learning successful. Learners require some amount of integration of all of the information being provided in the learning environment so that it makes sense and has meaning and utility in their lives. Company intranets permit employees to access issues facing the organization that can enrich the content of learning activities.

Accuracy and relevance: Information shared through e-learning should be accurate and appropriate. Instructors should employ measures (not only by direct contact with learners, but also by assessments such as surveys) to ensure that the content provided in the course is appropriate to learning needs. The course content should be reviewed regularly to ensure accuracy and relevance.

Several considerations must be taken into account for e-learning to be a beneficial investment and an effective knowledge management tool. The elements of the e-learning planning process include assessing and preparing organizational readiness (factors to consider before going online), determining the appropriate content (content that ties into the goals of knowledge management), determining the appropriate presentation modes (considering factors contributing to effective e-learning), and implementing e-learning (content and technology infrastructure considerations).

In academic institutions, ideally, both students and faculty should be provided with an e-learning environment that is optimal for each to be inspired to do the best job possible. For the student, the objective is to accumulate and learn as much knowledge as their personal make up will allow. The faculty must be stimulated and inspired to be outstanding faculty members for the students and simultaneously to grow professionally as rapidly as their personal abilities will allow. This is a complex environment to build as each individual involved will have different needs and hence different emphasis need to be placed on various components that make up the environment. No doubt it is impossible to provide the optimal e-learning environment for all individuals among the students and the faculty. It is important, however, to give considerable attention to this problem and to build the best e-learning environment possible with the financial resources available to the academic institution. Planning and faculty discussions make it possible to build e-learning and growing environments that are far better than would otherwise occur. E-learning although is a growing trend in the education community. Given all its merits, however, institutions considering this option should be aware of the particulars of WBT. With proper planning, implementation, and maintenance, you can create an optimal e-learning environment at your institution that benefits both the academic institution and students. In the following sections, different components and influences that impact on the magnitude of students and faculty's accumulation of and handling of knowledge is discussed.

TEACHING CULTURE

The development of a suitable teaching culture is a prerequisite of the academic world, once the principle to *have learned to learn* has been promoted. On the other hand, the development of the concept of sustainable employability determines that the university trains people to become future professionals with basic competences: cognitive, social, and affective, allowing them to achieve creative and effective professional performances in quick-change work environment (Oprean, 2006). Boyer states that, "... Without the teaching function, the continuity of knowledge will be broken and the store of human knowledge dangerously diminished" (Boyer, 1990). Teaching is not a process of transmitting knowledge to the student, but must be recognized as a process of continuous learning for both the lecturer and student. The old adage, if you want to know something teach it, certainly applies. But it needs to be extended (Al-Jumaily & Stonyer, 2000). The most obvious and instantaneous effect of development of suitable culture can be considered as improving the classroom environment and also classroom effectiveness. Through describing some important concepts the effective strategy for mutual relationship between learner and lecturer will be addressed.

Practical Applications

Completing a thriving course involves connecting the program of study to existent life problems and to present events. This is often made promising by including realistic applications in the classroom, and the instructor can carry out this by drawing on personal experience or by using student examples as sources of practical problems (Finelli, Klinger, & Budny, 2001). Students who contribute in certified work programs, research activity while in college can be a precious source of such information. Since there are many situations where bring students into contact with realistic problems, all that is necessary is to take advantage of those activities in teaching situations. Following paragraph give one example related to this partnership between industry and academic institute.

Faculty of engineering in Guilan University has a compulsory cooperative education program. Students are placed with an employer in their field of study (currently the school has partnerships with over 100 companies), and they alternate semesters at their worksite and on-campus beginning. Under this program, undergraduate students have some financial support from the university, and they get practical experience by working for the industry, for governmental agencies, or for public or private non-profit organizations. The connection with industry begins in the first year when students are teamed with alumni to assist the students with their semester writing projects. Through all of these experiences, students are often able to provide practical insight to classroom activities.

Learning Styles and Class Participation

It is extremely attracting and challenging area if one strikes a balance between lecturing and engaging in alternative teaching techniques to stimulate students with various learning styles. Learning style is a biologically and developmentally imposed set of personal characteristics that make some teaching (and learning) methods effective for certain students but ineffective for others. These include the Myers–Briggs Type Indicator, (Myers & McCaulley, 1985) Kolb's Learning Style Model, (Felder, 1993; Kolb, 1981) the Felder–Silverman Learning Style Model, (Dunn, 1990), and the Dunn and Dunn Learning Style Model (Dunn, Beaudry, & Klavas 1989). The following statements has been offered based on current research on learning styles to assure that every person has the opportunity to learn (Dunn, 1992):

1. Each person is unique, can learn, and has an individual learning style.
2. Individual learning styles should be acknowledged and respected.
3. Learning style is a function of heredity and experience, including strengths and limitations, and it develops individually over the life span.
4. Learning style is a combination of affective, cognitive, environmental, and physiological responses that characterize how a person learns.
5. Individual information processing is fundamental to a learning style and can be strengthened over time with intervention.

6. Learners are empowered by knowledge of their own and others' learning styles.

7. Effective curriculum and instruction are learning-style based and personalized to address and honor diversity.

8. Effective teachers continually monitor activities to ensure compatibility of instruction and evaluation with each individual's learning style strengths.

9. Teaching individuals through their learning style strengths improves their achievement, self-esteem, and attitude toward learning.

10. Every individual is entitled to counseling and instruction that responds to his/her style of learning.

11. A viable learning style model must be grounded in theoretical and applied research, periodically evaluated, and adapted to reflect the developing knowledge base.

12. Implementation of learning style practices must adhere to accepted standards of ethics.

An instructor who strives to understand his/her own learning style may also gain skill in the classroom. *Consider the question, how does the way you learn influence the way that you teach?:*

1. Most instructors tend to think that others see the world the way they do, but viewing things from a different learning perspective can be useful. It is good practice to specifically consider approaches to accommodate different learners, and this is often easiest after an instructor learns about his/her own learning style. An instructor with some understanding of differences in student learning styles has taken steps toward making teaching more productive (Finelli et al., 2001).

2. It is also important for the instructor to encourage class participation, but one must keep in mind the differences in student learning style when doing so. Research has shown that there are dominant learning characteristics involved in the perception of information through concrete versus abstract experience (Kolb, 1981; Kolb, 1984).

3. Some learners need to express their feelings, they seek personal meaning as they learn, and they desire personal interaction with the instructors, as well as with other students. *A characteristic question of this learning type is why?* This student desires and requests active verbal participation in the classroom. Other individuals, though, best obtain information through abstract conceptualization and active experimentation. This learner tends to respond well both to active learning opportunities with well-defined tasks and to trial-and-error learning in an environment that allows them to fail safely. These individuals like to test information, try things, take things apart, see how things work, and learn by doing. *A characteristic question of this learning type is how.* Thus, this student also desires active participation; however, hands-on activity is preferred over verbal interaction.

4. The instructor must have a sincere interest in the students. However, there is no best single way to encourage participation. Individual student differences in willingness to participate by asking questions often surface (Baron, 1998; Jung, 1971) Still, although the number of times an individual speaks up strongly depends on student personality qualities, a class where all are encouraged to enter into dialog is preferable, and opening the lecture to questions benefits all students.

Active Learning

An ancient proverb states, *Tell me, and I forget; Show me, and I remember; Involve me, and I understand.* This is the basis for active learning in the classroom, (Mamchur, 1990) and extensive research indicates that what people tend to remember is highly correlated with their level of involvement. It has been shown that students tend to remember only 20% of what they hear and 30% of what they see. However, by participating in a discussion or other active experience, retention may be increased to up to 90% (Dale, 1969).

TEACHING METHODOLOGY

Cooperative learning is a formalized active learning structure involves students working together in small groups to accomplish shared learning goals and to maximize their own and each other's learning. Their work indicates that students exhibit a higher level of individual achievement, develop more positive interpersonal relationships, and achieve greater levels of academic self-esteem when participating in a successful cooperative learning environment (Johnson, Johnson, & Smith, 1991; Johnson, Johnson, & Smith, 1992).

However, co-operation is more than being physically near other students, discussing material with other students, helping others, or sharing materials among a group, and instructors must be careful when implementing cooperative learning in the classroom. For a cooperative learning experience to be successful, it is essential that the following five elements be integrated into the activity (Johnson et al., 1991; Johnson et al., 1992):

1. *Positive interdependence:* Students perceive that they need each other in order to complete the group task.
2. *Face-to-face interaction:* Students promote each others' learning by helping, sharing, and encouraging efforts to learn. Students explain, discuss, and teach what they know to classmates. Groups are physically structured (i.e., around a small table) so that students sit and talk through each aspect of the assignment.
3. *Individual accountability:* Each student's performance is frequently assessed, and the results are given to the group and the individual. Giving an individual test to each student or randomly selecting one group member to give the answer accomplishes individual accountability.
4. *Interpersonal and small group skills:* Groups cannot function effectively if students do not have and use the required social skills. Collaborative skills include

leadership, decision-making, communication, trust building, and conflict management.

5. *Group processing:* Groups need specific time to discuss how well they are achieving their goals and maintaining effective working relationships among members. Group processing can be accomplished by asking students to complete such tasks as: (1) List at least three member actions that helped the group be successful, or (2) List one action that could be added to make the group even more successful tomorrow. Instructors also monitor the groups and give feedback on how well the groups are working together to the groups and the class as a whole.

When including cooperative learning in the classroom, the instructor should do so after careful planning. Also, the students may be more receptive to the experience if the instructor shares some thoughts about cooperative learning and the benefits to be gained by the activity.

In our classes, the students should be able to do group work; in fact much of the assessment takes the form of performance presentations in which two students from each group act as directors in a given week. The remaining group members are just helping, so it is important that they have reliable access to the directions in advance of each week's performance. The student's online collaboration can work on their joint presentations at a distance from each other. I have to oversee the group interactions to check that groups were not copying from each other and to ensure that directions were posted on time, so I helped to set that up.

It should be noted that there is a big difference from one module to another. So in modules where the tutor has told the students that they will log-on to the online discussion at such and such a time and they give replies and responses to the student's exchange—then you see the majority of students participating.

There is an interesting balance—the students want to know you are there, even if you are not directly participating in the discussion. We think the students want validation that the opinions they are expressing are appropriate—so if they feel that the tutor is in the background and will correct any major mistakes they seem to be more open to learn from each other.

In particular, the objective was to assess the students' ability to engage with the material in a certain way, rather than to have them reiterate a series of facts about the material. There is an acceptance across the whole school that e-learning is part of teaching now. Some students are happy to embrace it while others are still a bit anxious and have workload concerns. If a school does want a wide-scale adoption of e-learning there does need to be a school based individual who can coordinate and provide support for students? I could not correlate between team work and e-learning in this paragraph.

We think that our graduates should also be provided with an international platform as a foundation for their careers. With this in mind, language skills are essential for engineers who are operating in the increasingly global environment. Therefore, it is necessary to speak one or two foreign languages in order to be able to compete internationally. It is also necessary to include international subjects in an institution's programs of study; to make programs attractive to international students.

Nevertheless, engineering education institutions still grapple with the fundamental concepts and ideas related to the internationalization of their activities and courses. Comprehensive studies concerning curriculum development and its methodology are essential in order to ensure that the main stream of academic activities is not completely lost in the process of globalization. Research should be undertaken on a global engineering education curriculum, in order to identify fundamental issues and concerns in an attempt to devise and develop a proper methodology, which would be used in curriculum development in an era of globalization.

TEACHING INSTRUMENTS

It is widely believed that using traditional methods like writing of teaching material on whiteboard or blackboards take lots of time and cause tutor contact diminished to lowest level. Nowadays, fortunately, majority of universities are equipped with audio–video learning devices to communicate with students. Using electronic devices like opaque, overhead, and video projectors take the lowest time for write up of material in the class, and attract student's attention to completely focus on matter without wasting time. Also there is a direct eye contact between tutor and learner which is necessary to oblige student to be in class without any absence of mind. It is necessary that all material would be transferred as hard copy in advance to student to completely use an interactive environment and do not coerce to write teaching material in class. Student goes ahead with tutor and main part of understanding process is completed in class.

DISCUSSION

The economic future within Europe and worldwide increasingly depends on our ability to continue to provide improved living standards, which in turn depends on our ability to add value to products and services continuously. To achieve this, employers increasingly need to look to engineers throughout their organizations to come forward with innovative and creative ideas for improving the way business performs and equally to take greater responsibility within their work areas through personal development on a life-long basis.

To compete in the rapidly changing global markets it has become essential for organizations to recognize that they can no longer depend on educated elite but, increasingly, must maximize the potential of the workforce within the organization by harnessing all available brain power to assure economic survival and success. Life-long learning will provide for the development of a more cohesive society, much more able to operate within a cross-cultural diversity.

We expect steady growth in global engineering e-learning programs. The main drivers will be the strong industry interest in recruiting students who have some understanding and experience of global industry, government funding agencies who are slowly developing support for global engineering education, rising faculty awareness, and high levels of student interest. It is not, however, easy to do, and resource issues will slow the growth rate. The following are the reasons why we think global engineering e-learning programs are good:

- Preparing students for the global economy. This is necessary and it will happen.
- Everyone learning from the comparative method: education, research, and service global collaborations make everyone smarter.
- There are good research prospects through the education activities such as optimizing virtual global teams, and through research collaborations that are a byproduct of the educational collaborations.
- It builds international and cross-cultural tolerance and understanding.

CONCLUDING REMARKS

It is important to emphasize, the fact that learning technology standards implement a certain level of interoperability. In order to achieve the smooth co-operation of all e-learning components we should impose standards in every procedure. Standardization committees should define standards that cover all aspects of the educational procedure and do not cover each other.

A major complaint about e-learning standards is that products claiming conformance do not work together without further tweaking. This translates into lost time and expensive service engagements. As a result of this challenge, there is an increasing emphasis on developing conformance tests and certification programs. It is necessary that e-learning standards must be adopted by everyone without any customization or modification (i.e., based on differences in language, country, low, customs, etc.). The roadmap to achieve standardization of e-learning technologies comprises the following steps:

1. First, we should overview the e-learning process as a whole. We must define the operations included in the e-learning process, the information exchanged (input, results, etc.). In this step, we should stabilize the existing practices and record the existing standards and needs.
2. The second step is to locate the main standardization bodies and have them work for the common aim. International boards must decide on the standards by taking into account the needs.
3. The third step concerns the thoughtful definition of specifications. The specifications should cover all possible needs of e-learning systems and avoid redundancies.
4. The final step comprises the dissemination of specifications and their stabilization into standards. Once they are defined, the specifications are communicated to the community for testing. Additional requirements or modifications.

One key finding was that the students wanted a greater sense of community; they wanted more interaction with lecturers and the university. Many professors post a course syllabus, homework assignments, and study guides on the Web, and some ambitious faculty with large classes may even give exams online. But educators are not ready to plunge completely into the e-learning environment and fully adopt simulated lab experiments or self-guided online instruction. One reason is the time it would take for faculty to learn how to use new computer programs and to develop online materials to replace their current versions. Another reason is that laboratory science requires intuitive observations and a set of skills learned by hands-on experience, none of which can be fully imitated in the digital realm. Simulations are a bridge from the abstract to the real, combining old technology with new technology to connect theory in the classroom with real-world experience in the lab. The purpose is to provide a creative environment to reinforce or enhance traditional learning, not to replace it. As students progress through the theory section, they can quiz themselves on what they have learned. As an example, in a virtual lab section, students learn how to construct a data table by using chemical shifts and coupling patterns from spectra of common compounds. Errors entered by the students are automatically corrected and highlighted in red. Once the table is complete, the student is directed to select the molecular fragment associated with an NMR signal, and then assemble the fragments to form the molecule.

E-learning has great potential for engineering education. Broader use of e-learning will be driven by the next generation of students who will have had exposure to e-learning programs in high school and will start to ask for similar systems at the undergraduate level. E-learning also will likely be adopted more quickly for distance-learning courses or courses for which lab costs or lack of lab facilities may be a factor, such as those for nonscience majors or those offered by community colleges or high schools. It should be noted that the e-learning model adopted in one university cannot be the best model to follow in another college. The providers of distance learning may have to accept that there are limitations in all models of distance education. The best opportunity lies in identifying and offering the mode that suits the most students in a particular cultural and regional context at a particular point in time. There is still work to be done on distance learning, but initial signs are positive. For all modes of delivery, learning is the active descriptor, distance is secondary. Nevertheless, the Website appears to be an effective supplementary tool for students with all learning style modalities. The correlations between course grade and the Website usage were weakest for active and sensing students. It may be an issue that needs to be addressed through instructional design to make the materials more engaging for these particular modalities. Also, the relatively small sample of styles may have affected the results.

The adoption of standards and specifications facilitates the dominance of platform independent, open technologies, and promotes user-centric e-learning systems. Standardized technologies have several merits that protect and nurture an e-learning investment. These are in general:

- *Interoperability:* Content from multiple providers can be easily disseminated within consumers and a multitude of systems. Problems of translation, communication, and information exchange are easily solved and transparent interaction of systems is achieved.
- *Re-usability:* Content and code can be assembled, disassembled, and re-used quickly and easily. Additionally content objects can be adapted and used in a context other than that originally designed.

- *Manageability:* Systems can track the appropriate information about the learner and the content. Learners' profiles, educational target, and content "speak the same language" so it is easier to find, manage, and assembly "the right stuff" for each case.
- *Accessibility:* A learner can access the appropriate content at the appropriate time on the appropriate device. Content warehouses can be developed and become available to amateurs or professionals that use any application based on the common standards.
- *Durability:* Content is produced once and transplanted many times in different platforms and systems with minimum effort. Buyers are not "trapped" by a particular vendor's proprietary learning technology and their investments become permanent and adverse to risk.
- *Scalability:* Learning technologies can be expanded in functionality in order to serve broader populations and organizational purposes. An organization's return on investment in e-learning products can increase if they can be leveraged beyond their original scope.

The standardization of procedures in e-learning can be achieved through the cooperation of all participants of the e-learning community: developers, vendors, and users should work together in order to create, validate, establish, and disseminate standards for every e-learning task.

KEYWORDS

- **E-development**
- **E-learning**
- **Learning Style Model**
- **Software tools**
- **Web-based training**

Chapter 5

Web-Based Training: Design and Implementation Issues

B. Noroozi and A. K. Haghi

INTRODUCTION

Web-based training (WBT) is an innovative approach to distance learning in which computer-based training (CBT) is transformed by the technologies and methodologies of the World Wide Web, the Internet, and intranets. WBT presents *live* content, as fresh as the moment and modified at will, in a structure allowing self-directed, self-paced instruction in any topic. WBT is media-rich training fully capable of evaluation, adaptation, and remediation, all independent of computer platform.

WBT is an ideal vehicle for delivering training to individuals anywhere in the world at any time. Advances in computer network technology and improvements in bandwidth will usher in capabilities for unlimited multimedia access. Web browsers that support 3D Virtual Reality, animation, interactions, chat and conferencing, and real-time audio and video will offer unparalleled training opportunities. With the tools at hand today, we can craft highly effective WBT to meet the training needs of a diverse population. Web-based performance support systems (WBPSS) further help today's busy workers perform their jobs by integrating WBT, information systems, and job aids into unified systems available *on demand*.

The current focus of WBT development is on learning how to use the available tools and organize content into well-crafted teaching systems. Training designers are still struggling with issues of user interface design and programming for high levels of interaction. Unfortunately, there are few examples of good WBT design visible on the public Internet. As instructional designers and training analysts learn how to write and produce WBT, and as training vendors come to realize the overwhelming advantages of this delivery method, expect an explosion in training offerings available over the public Internet and private intranets.

The virtual environment of e-learning courses can provide cheaper, safer, more comprehensive and more inclusive approaches to engineering educational material. What usually the students need to learn in laboratories and workshops and it is costive and demanding for the universities and the schools. The aim of this chapter is to count the requirements of engineering education and to accord the facilities and inadequacies of e-learning as training technique in engineering instruction.

Instructors of online courses perceive interaction as an important aspect of a successful learning. A number of technologies and instructional activities are employed to promote course interactions in general. However, instructors vary when using the technologies and instructional activities that require sophisticated technical skills or that need a significant amount of time to learn how to use them effectively. Furthermore, instructors admit that they have difficulty making their online courses as interactive as they wish. The difficulties of reaching a desired level of interaction in online courses appears not only is associated with faculty limitations in their technical skill and/or lack of time, but also is related to "old habits" or mind sets that they have developed over many years of teaching in traditional face-to-face educational settings. When the teaching context changes to a totally new environment, instructors face difficulties transforming their instructional skills that have been accumulated over the years. On the other hand, it is exciting to observe that some creative instructors attempting new instructional activities and technologies that are unique to online education. To help reduce such gaps of teaching skill in online environments, periodic experience sharing among online instructors, experts, and instructional specialists is recommended.

While interaction, in all its varied formats, is perceived as an effective means for learning, students tend to vary in their preferences about additional interaction in their online courses. Such variations tend to be related to individual personalities or learning style differences. Further, research is needed to determine the relationships between learner preferences related to online interactions and individual differences. Although, not everyone hopes to have highly interactive online courses, lowered expectations do not necessarily mean that they do not want higher interactions. Therefore, instructors should continue to search for effective instructional strategies to overcome the numerous barriers of interactive online learning environments.

Distance learning is a progressive method of pedagogy employing the electronic technology to transfer the education to the students who are not on site and allows them to participate in educational activities on their own time. Since 1728, which the first ideas of distance learning has flickered through application of postal services, this technique has passed through several transition states like radio, television, telephone, computer, and internet. Emersion of internet made a revolution in distance learning leading to a new generation called web-based learning. Web-based learning is practically the milestone of electronic learning or e-learning, profiting the popularity of the internet as a fast information source.

Connolly and Stansfield (2007) have categorized the historical progress of WBT through three distinct generations. The first generation took place from 1994 to 1999 and was marked by a passive use of the Internet where traditional materials were simply repurposed to an online format. The second generation appeared through 2000–2003. This spectrum is marked by the transition to higher bandwidths, rich streaming media, increased resources, and the move to create virtual learning environments that integrated access to course materials, communications, and student services. Internet, Intranets, local area networks (LAN) and wide area networks (WAN) have offered the learners the opportunity to use distance learning beyond pre-recorded classes, educational software, and virtual laboratories. The third generation is currently in progress and it is marked by the incorporation of greater collaboration, socialization, project based learning, and reflective practices, through e-portfolios, wikis, blogs, social bookmarking and networking, and online simulations. Online forums, blogs and discussion boards have become a precious resource in Learning Management Systems

(LMSs). They allow learners to communicate with their compeers and tutors, empowering them to socialize and learn together online. Furthermore, the third generation is progressively being influenced by advances in mobile communicating (Connolly, 2007; Harper, 2004; Monahan, 2008).

Due to its singular capabilities, web-based learning has entered and is widely used in every field of science and technology. It is not only warmly welcome at schools and universities, but also in factories and houses. However, utilization of web-based learning technique requires tender and comprehensive attentions in designing, applying, and assessing configurations, directed by when, where, and for which purpose it is being employed.

Among different branches of human knowledge and sciences, engineering, as well as medicine is more involved in practical and daily-life aspects, where the virtual utilities and educational software can be utilized to consummate the practical features of engineering education. Furthermore, the virtual environment of e-learning courses can provide cheaper, safer, more comprehensive, and more inclusive approaches to engineering educational material. What usually the students need to learn in laboratories and workshops and it is costive and demanding for the universities and the schools. The aim of this chapter is to count the requirements of engineering education and to accord the facilities and inadequacies of e-learning as training technique in engineering instruction.

ENGINEERING

There is a growing trend among medical educators towards the use of new learner-centered teaching and preparation methods, based on self-learning and with specific objectives. In contrast to conventional theory lectures, they are more efficient in promoting learning, are more flexible both for teacher and pupil, and moreover, they help the learner to acquire the self learning habit, which should become a daily practice over the course of the learner's professional life. In relation to computer aid and web-based learning, there is extensive literature that shows how the computer is effective in the instruction of engineering professionals in comparison with conventional education, especially programs which include problem-solving or interactive methods. In engineering self-education it facilitates the learner's attention, allows individualized progress and provides immediate non-competitive and flexible feedback, adapted to individual needs. Engineering is the discipline of implementing of science and mathematics to develop explanation for the problems and to find practical solutions that have an applicable outcome. Engineering is the knowledge of creating a new world based on the rules which science has already discovered. Engineers suppose to be inventors. They design and manufacture machines, processes, systems, and even economical constitutions. Utilizing science and mathematics, they observe the world, find the troubles, dream up new ideas and realize the dreams to improve the quality of life and make our world a more comfortable place to live. Based on scientific principles, every engineering discipline has two main tasks (Wintermantel, 1999):

- To model subsystems using the theoretical and methodological scientific knowledge. In this respect, engineering is not different from the natural sciences.

- To develop methods and procedures, which allow real systems in all their complexity to be designed and constructed even if not all subsystems have been precisely modeled due to a lack of a thorough knowledge of the underlying physics and chemistry.

The Engineers Council for Professional Development (ECPD), in USA has defined the term of engineering as:

> the creative application of scientific principles to design or develop structures, machines, apparatus, or manufacturing processes, or works utilizing them singly or in combination; or to construct or operate the same with full cognizance of their design; or to forecast their behavior under specific operating conditions; all as respects an intended function, economics of operation and safety to life and property. (Science, 1941)

Requirements to Make the Best Engineer

An engineer supposes to solve complex problems by the simplest solutions. They are often responsible for directly creating a new product or service. The main functions of an engineer have been defined through seven terms by Encyclopedia Britannica (Smith, 2009):

1. *Research*: Using mathematical and scientific concepts, experimental techniques, and inductive reasoning, the research engineer seeks new principles and processes.

2. *Development*: Development engineers apply the results of research to useful purposes. Creative application of new knowledge may result in a working model of a new electrical circuit, a chemical process, or an industrial machine.

3. *Design*: In designing a structure or a product, the engineer selects methods, specifies materials, and determines shapes to satisfy technical requirements and to meet performance specifications.

4. *Construction*: The construction engineer is responsible for preparing the site, determining procedures that will economically and safely yield the desired quality, directing the placement of materials, and organizing the personnel and equipment.

5. *Production*: Plant layout and equipment selection are the responsibility of the production engineer, who chooses processes and tools, integrates the flow of materials and components, and provides for testing and inspection.

6. *Operation*: The operating engineer controls machines, plants, and organizations providing power, transportation, and communication; determines procedures; and supervises personnel to obtain reliable and economic operation of complex equipment.

7. *Management and other functions*: In some countries and industries, engineers analyze customers' requirements, recommend units to satisfy needs economically, and resolve related problems.

By integrating all above mentioned functions, the required skills for all engineering disciplines can be sorted in 3 categories: mathematics and science, team working and curiosity, creativity and innovation (Figure 5.1). Every educational curriculum for engineering education has to fulfill these themes comprehensively.

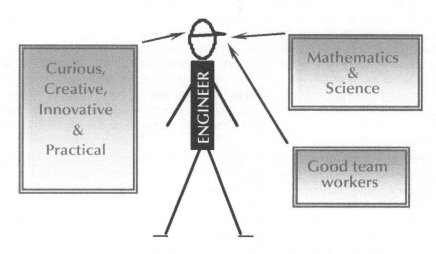

Figure 5.1. The essential requirements of being an engineer.

Mathematics and science: Math and science are the basis of engineering. Physics, chemistry, biology, economics, psychology as the laws, forces and resources of nature are the primal necessities of different engineering disciplines. The physical sciences are concerned with the laws that govern the world and the realities of nature. Mathematic however is a comprehensive concept that, in spite of being inspired by the rules of nature, does not bound up with it. It can exist and grow within itself. Mathematics and basic sciences are pure. Engineering, in contrast, is not pure. The process of training an engineer to apply the basic rules of science and to design a machine or a process requires a study of engineering models and deep understanding of basic science and the mathematical techniques. Those models require correspondence with reality in their conception, and precision in their description. And those mathematical techniques, like all mathematical techniques, require practice, sophistication, and rigor. Consequently, the technological world of an engineer has been constructed on the pure disciplines of mathematics and basic sciences, but it is not contained in them (Chatterjee, 2005).

Teamwork: Most engineers work as a team. Therefore they should be good team players and acquire reasonable skills in communicating with others. This is highly important when planning and creating new projects. Teamwork and group work are increasingly common elements of engineering learning. When group activities and projects are assigned, two types of learning outcomes will emerge:

1. *Product outcomes*: outcomes related to the content of the course;
2. *Process outcomes*: outcomes related to team skills and contribution in group work activities.

The aims of teamwork development are that the members work interdependently and work towards both personal and team goals. They understand these goals are accomplished best by communal support. They feel a sense of ownership towards their role in the group because they committed themselves to goals. Members work in partnership together and use their talent and experience to supply the success of the team's purposes. As a part of every mutual work, members try to be honest, respectful, and listen to others point of view. They see conflict as a part of human nature and they react to it by treating it as an occasion to hear about new ideas and opinions. Members participate equally in decision-making, but each member understands that the leader might need to make the final decision if the team cannot come to a common consent (Kaufman, 2000).

Practical, Innovative personality: Conceptual advances have always been the driving force behind scientific progresses. This in turn relies on creativity and the ability to continue the production of new insights and innovative ideas. The history has given us many discoveries (especially in engineering and the sciences) linked to a sudden recognition, which are the outcomes of creative, dreamer mind of human beings. An engineer is who dreams, then assesses the realization possibilities of the dream and finally makes the dream a reality. Therefore, an engineering student on one hand should have the talent and on the other hand has to be trained up appropriately to be able to follow the steps of an engineering decision according to Figure 5.2 and to create and find the answers for questions like:

- What is the problem?
- What do I want to do?
- What are the benefits of this solution? And are they significant?
- How can I implement the solution?
- How can I know that if I have reached the benefits?
- How can I represent a better solution with my new experiences?

ENGINEERING EDUCATION REQUIREMENTS

In a continuously changing world, many challenges and opportunities face the engineering profession and engineering education. Based on various technological and economic developments, fundamental changes have taken place over the last few years that affect the methods in engineering education. In consequence, the actual training trends like the trend to virtual and classical Instructional methods, the global availability of information on one hand and of information resources and accessibility on the other hand, the rise of learning demands and the particularities of engineering education require a strategic streamlining of each of these options. The basis for this strategic planning should be the instructional principal abilities and the use of the general available enabling technologies.

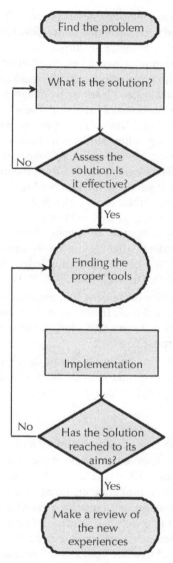

Figure 5.2. Decision-making in engineering processes.

Engineering professionals working in industry are in general unsatisfied with the level of real-world preparedness acquired by recent university graduates entering the work market. Their dissatisfaction is understandable. In order for these graduates to be productive in an industrial setting, one idea is that the organizations which hire them must supplement their university education with extensive on-the-job training and preparation that provides them with the skills and knowledge they lack (Baker, 2005, Conn, 2002).

On the other hand, following the tremendous progress in science and technology, engineering teaching materials are currently under development by the multidisciplinary (production engineers, computer scientists, economists) project teams. The philosophy behind this approach is to give access to all information that is relevant to understand and to work on the topic of interest. The background of the didactic approach of these projects is teaching and learning by distance instructions. The central focus is the specification of event-oriented lessons and seminars such as computer supported role-plays for distributed problem-solving, gaming in an open environment and multinational web-based teaching, where students will learn to work on open problems in different fields of engineering, having an active connection with other universities and also the manufacturers. They will be guided from theoretical learning in lessons via interdisciplinary distributed games to international teamwork in cooperation with partner universities. Step by step, their capabilities and entrepreneurial spirit to apply theoretical knowledge will increase, as well as their social and managerial skills. This approach takes up the idea of e-learning to enhance traditional more or less passive reception of knowledge by students what we know as learning by listening, towards an active, self-directed learning, and cooperative knowledge creation learning by trying (Baker, 2005; Grunbacher, 2007; Hirsch, 1998; Pyrz, 2003).

Concerning to the necessities of modern life, engineering education requires essential transitions in methods and tools to accord with modern societies. So that, engineering education designers endeavor to design and to employ the most consistent methods which fulfill this purpose.

As a potent approach, WBT technique is one of the modern methods of distance learning which plays an advancing role in every field of education. It can be applied solely or joined with the traditional education systems.

WEB-BASED TRAINING

In order to overcome the problems of incompatibility and employ a new way of linking made possible by computers, called hypertext, a group of scientists at the European Laboratory for Particle Physics (CERN) in Geneva, Switzerland, began developing an Internet tool in 1989 to link the information achieved by all of the CERN researchers. Rather than presenting information in a linear or hierarchical fashion, hypertext permits information to be linked in a web-like structure. The tool provided a way to link textual information on different computers and created by different scientists. Nodes of information could be linked to other nodes of information in multiple ways. As a result, users can dynamically interlace the information in order to make them most convenient. The CERN project resulted in an innovative front-end to the Internet, now called the World Wide Web (WWW) (Gottschalk, 2009).

For educators, the Internet and HTML have provided a new occasion for distance training. The ability of putting together graphics, text, sound, animations, and movies into a single tool has empowered the educators to build virtual classroom on their WebPages. By this mean, the organizations or individuals can create their own home pages and link them to other home pages or share them with other computers. The educational home pages can be supported by virtual laboratories, virtual libraries, text lessons, virtual classrooms using video conferencing, exercises, and exams. These are all the body builders of WBT.

As a simple definition, WBT is a form of training that can be developed by utilizing a network.

Web-Based Training (WBT) is the delivery of educational content via a Web browser over the public Internet, a private intranet, or an extranet. Web-based training often provides links to other learning resources such as references, e-mail, bulletin boards, and discussion groups. WBT also may include a facilitator who can provide course guidelines, manage discussion boards, deliver lectures, and so forth. When used with a facilitator, WBT offers some advantages of instructor-led training while also retaining the advantages of computer-based training. (Imperial-College-London, 2009)

WBT is the innovative approach to distance learning in which training material is transformed by the technologies and attitudes of the Internet and intranets. The specification of WBT is the presenting of live contents, in a structure allowing self-directed, self-paced instruction in any topic (Kilby, 2009). The same as all other distance learning techniques, WBT needs some indispensable components which can be categorized in three groups:

1. Instructional design
2. Knowledge delivery techniques
3. Standards

Specifications and examples of these components are illustrated in Table 5.1 (Bergstrom, 2001; DCMI, 2009; Digitalthink, 2003; Friesen, 2004; IEEE, 2005; IMS 2009; Le@rningFederation, 2008; Ryder, 2008). These are general concepts, details of which need to be revised and upgraded according to the demands in different educational fields.

Table 5.1. Specifications and examples of indispensable components.

Concept	Aspects	Examples
Standards	Resource detection	• ADL SCORM Content Aggregation Model • IEEE P1484.11 Computer Managed Instruction • IMS Content Packaging Specification • IMS Content Packaging Specification • IMS Simple Sequencing Specification
	Content management	• Dublin Core Metadata Element Set • UK LOM Core • IEEE 1484.12.1-2002 Learning Object Meta-data standard • IMS Learning Resource Meta-data Specification • EdNA Metadata Standard • Gateway to Educational Materials Element Set • Canadian Core Learning Resource Metadata Application Profile • SingCore

Table 5.1. *(Continued)*

Concept	Aspects	Examples
	Accessibility	• Le@rning Federation Accessibility Specification • Web Content Accessibility Guidelines • IMS AccessForAll Meta-data Specification
	Interoperability	• ADL SCORM • IMS Resource List Interoperability Specification • IMS Enterprise Information Model • IMS Enterprise Services Specification • IMS Question & Test Interoperability Specification • IMS Shareable State Persistence Specification • IMS Digital Repositories • IMS Learner Information Package Specification • ISO/IEC JTC1 SC36/WG3 Learner Information
	Pedagogy quality	• Le@rning Federation Educational Soundness Specification
	Digital rights management	• SO/IEC JTC1 SC36/WG4 Digital Rights Expression Language • Le@rning Federation Rights Management Specification
Instructional design	Analysis Design Development Implementation Evaluation	• ADDIE • ASSURE • ARCS • CSCL • CSILE
	Synchronous technologies:	• Telephone and Mobile • Web-based conference • Video conference
Knowledge delivery techniques	Asynchronous technologies:	• Postal materials • Audio and video cassettes • Radio and Television • Fax and E-mail • Voice mail, MMS and SMS • Message boards and Forums • E-portfolios and blogs

Web-based Training for Engineering Students

Engineering education is in fact engineering training, explained as the way of improving a person's performance related to job environments by providing instructions. Intention, designing, tools and media, and assessment are usually the components of training programs which can increase the effectiveness of the Engineering training. Though, intent is the motivation and willing to participate in parts of the training, the design is the systematic structure of the training. It determines each step while the learners go ahead with the new skills. It also includes instructional strategies and related measurement issues about these strategies. Tools and media are the means of instruction transfer. It could be a classroom, a WBT system, or combinations of different tools. To finalize the learning process, a formalized assessments and certification provide accountability of the training (YilDirim, 2009).

Two basic components of WBT technology are Ideal Learners and Ideal Instructions. These are the delicate concepts of WBT needing to be taken care for engineering education. As Horton states an Ideal Learner is who (Horton, 2001):

- Has definite goals;
- Opens to learn independently and view learning prohibitively;
- Is self-disciplined and have good time management;
- Is moderately capable of field and know basic concepts and facts;
- Has enough good writing skills;
- Has enough domination in computing skills;
- Has positive attitude toward technology in business and learning; and
- Is so relaxed while there is any technical problem and can cope with these problems.

An Ideal Engineering Learner in WBT system, in addition to above mentioned items has to have notable skills in mathematics be able to adapt himself with the group working and the projects which are usually defined as part of the learners assessment plan and schedule of the course.

The most important and crucial aspect to reach to an advantageous ideal course is determining learners needs. In an ideal course, teaching is efficient. Precise objects are well defined and the learner acquires new knowledge by spending less time. Development process of WBT is the other crucial concern. Using web-based utilities, the designers can use several models and modulus for instructional objects.

The nature of WBT provides a remarkable array of opportunities and technologies such as XML (eXtensible Markup Language), Web Services, Peer-to-Peer, and many other approaches. On the other hand, there are many items in WBT environment influencing the outcome. Therefore one can find the most and the least useful information simultaneously. Hence, the designers of WBT courses are responsible of performing the course in the most comprehensive and specific system. Michael Allen counts three priorities for successful ideal course designing (Allen, 2003):

1. *Motivation*: Motivation is essential to learning because it induces attention, persistence, and participation in learning activities. This concept is very familiar to literature of learning and instructional designers, but unfortunately, explicit managing of motivation seems to be creeping into educational plans very slowly. The fact is that, highly motivated learners will find the means and the tools to access the learning material. Motivation is definitely a very effective task in engineering education, because it is directly connected to creativity

and curiosity. A designed approach of motivation in engineering courses assists team working approaches and can encourage the students to exchange their knowledge more readily.

2. *Directing*: Highly motivated learners are ready to use any tool helping them to reach the learning material. Therefore, it is essentially important to design and provide appropriate educational materials upon a precise schedule. By this way the effectiveness of motivation credit will not be spoiled. Priming appropriate learning content includes providing excellent indexing helping the students to identify an appropriate learning material, or applying an assessment that specifies what the students (in more progressed educational courses, what every individual student) may need. For engineering aspects the relevant directions are even more important. The engineering aspects are widespread in different fields of science and technologies. Meanwhile being founded on basic sciences concepts; it has more connection with applied science. Engineering is in fact the most notable representative of applied science, and due to the broadness and variety of its subjects it requires thoroughly disciplined plan for courses.

3. *Purposeful and impressive*: An ideal course has to be designed for the wide spectrum of students with different capabilities. It has to be enough understandable for less skilled learners meanwhile it is privileging talent students to grasp deeper concepts with satisfactory and appealing supplements. However, from another point of view providing appealing and purposeful material for the courses impresses the students and helps them to memorize and remember the instructional contents more easily.

For engineering courses designing this approach is even more desirable and helpful, since as objective concepts, the engineering perceptions can be significantly described using the virtual aids of WBT. The main advantage of this approach is the ease of remembering the taught objects in real job status. To facilitate the instructional course with such configuration, the WBT designer may use the following utilities:

- Attractive context and new examples;
- Providing a charming environment using several means represented by virtual world advantages, such as computer simulated environments;
- Problem-solving scenarios;
- Engaging themes, media, and user interface elements; and
- Practices, that in WBT cases Virtual Laboratories may empower the instructional contents with virtual practical sessions.

To express the advantages of the above mentioned issues, some simple but very applicable examples are presented in the following paragraphs, revealing the application of WBT in some engineering subjects.

Cases

Thermodynamic is the fundamental knowledge of many branches of engineering such as chemical engineering, mechanical engineering, textile engineering, polymer engineering, and many others. One of the fundamental lessons taught in this course is

Carnot cycle to make an example of reversible heat engine. This cycle is defined by several adiabatic and isothermal steps including several changes in volume, pressure, enthalpy, and entropy of the system. It is a fairly complicated introduction on second law of thermodynamic inaugurating the headwords for more elaborate concepts. As a pivotal subject, teachers usually have to put a lot of time on it and even in this case the students look still confused.

Professor Michael Flower from Department of physics, University of Virginia, has an online course on modern physics, entitled "Galileo and Einstein" at http://galileoandeinstein.physics.virginia.edu, contenting different fundamental laws of physics such as Galileo's Compound Motion, Kepler's law, Doppler Effect, Carnot Cycle, and many others. The advantage of his online course is simplification of subjects providing animations, graphs, definitions and states, and all these are connected to each other. Through the visual and virtual interface of this course, the student would be able to follow easily the instructions. They would have the opportunity to make the calculations on each point of cycle, and view the correspondent effects on states, graphs, and the heat engine status which they have to learn. Moreover, it is easier for the teachers to teach (Flowers, 2008; Hwang, 2004; Ion, 2007).

Polymer chemistry is the fairly new subdivision of chemistry being taught in chemical engineering, polymer engineering, fiber production engineering, textile engineering, composite and material scientists. Polymer science which is in fact the chemistry, rheology, and processing of macromolecules has intricate subjects for both students (to learn) and professors (to make sure that the taught subjects have been understood by the students). American chemical society (ACS) has several short courses on chemistry and chemical engineering, including polymer courses. The ACS short courses in polymer chemistry targets industry professionals, chemists, materials scientists, and engineers with a B.S. or higher degree seeking greater understanding of principles governing polymer science and a comprehensive, up-to-date overview of polymer subjects. Since no assumptions are made regarding the background of the participants, the courses attempt to address a variety of learning styles and knowledge bases through lecture, demonstrations, laboratory exercises, and computer-based independent study (ACS, 2009).

The department of polymer science of the University of Southern Mississippi has established an online center, called Polymer Science Learning Center (PSLC). This center contents polymer courses for different levels from beginners, even kids to polymer explorers. As a multi-level, multimedia polymer education resource, PSLC is an interactive website in polymer education by pioneering an innovative hands-on, inquiry-based, multi-faceted learning environment for pre-kindergarten through researchers that will result in all polymer education ventures originating by choice with the PSLC. The mission is to teach polymer science to students of all levels in the most memorable manner possible to better prepare them for both chemical careers and life (Mathias, 2009).

CONCLUSION

Though interaction is often billed as a significant component of successful online learning, empirical evidence of its importance, as well as practical guidance or specific interaction techniques continue to be lacking. In response, this study utilizes both quantitative and qualitative data to investigate how instructors and students perceive the importance of online interaction and which instructional techniques enhance those interactions. Results show that instructors perceive the learner-instructor and learner-learner interactions as key factors in high quality online programs. While online students generally perceive interaction as an effective means of learning, they vary with regard to having more interaction in online courses. Such variations seem to be associated with differences in personality or learning style. The present study also shows that instructors tend to use technologies and instructional activities that they are familiar with or have relied on in traditional classroom settings. When it comes to learning more sophisticated technologies or techniques, instructors vary significantly in their usage of new *approaches*. World Wide Web and Internet technologies are the main support of modern communication. They have established foundations of modern educational systems which are offering an extraordinary volume of information to a wide spectrum of learners, even at schools, universities and factories, or house users which are looking for specific objects. Internet contains some of the vantage components of distance education naturally supplying time and place independency as the server computers are working 24 hours a day, all over the year.

With all the advantages happening in distance communication, the teaching and training technology is gravitating more and more toward web-based aspects. The main users of web-based training technologies are still the students of the universities and schools. However, due to its virtual supplements, WBT can start more effectively in some particular branches of science such as engineering.

WBT is a novel instructional technique offering many superior advantages over traditional methods. Potential capabilities of WBT nominate it as a suitable method for engineering education, some of which can be counted as:

- WBT offers time-effective and cost-effective training services.
- WBT overcomes the disadvantages of traditional instructional methods based on printed course material. In WBT the educational content can be changed or added fast and easily.
- WBT provides standardized courses ensuring congruous training.
- The educational content can be easily designed to be consistent with different budgets and cultures.
- Course material can be easily adapted and updated for different learners and according to their needs.
- Virtual environment of WBT provides very powerful configurations that can replace the costive practical sessions required in engineering education.

In addition to all the advantages of WBT, it includes some disadvantages as well:

Initial implementation of a WBT system can be more costive than the cost of publishing course material as textbooks.

- Bandwidth limitations and other limitations on required electronic devices such as the capabilities of computers are considered as the constrains of WBT development.
- WBT is more adapted with self-directed students. But the students who are more used to traditional techniques of teacher- classroom - printed textbooks or the students who do not have enough skills in using computer or internet may find it difficult to get adapted.
- WBT provides time and place independency for the students; however, the students are more possible to get interrupted because of the freedom they gain because of the natural self-dependency of WBT method.
- Even with the existence of Virtual laboratories, the students will not get the idea of the real training, as well as they are needed to be trained for their coming careers.

Collecting all the presented information, WBT can be counted as a powerful method when it is designed and applied in its right position. It is certainly the most perfect method to enter in the course content and actually start it, especially for engineering courses where the provided virtual environment of WBT can help the students to figure out the subject. WBT would be very effective when the technical capabilities of the students and their skills have been depicted realistically. The main advantage of WBT is that it can be easily customized regarding to the needs. The needs can be the level of education, the required skills, job aids, the culture of users, and many other issues.

KEYWORDS

- **American chemical society**
- **Computer-based training**
- **Virtual environment**
- **Web-based performance support systems**
- **World Wide Web**

Chapter 6

Teaching Qualitative Research Methods to Students 2.0: A Reflexive Arts-Based Approach

Rocci Luppicini and Jessica Backen

INTRODUCTION

Arts-based inquiry and reflective practice are innovate research approaches used within educational contexts to help individuals articulate profound aspects of their learning and identity. This chapter focuses on the use of a strategic application of arts-based reflective practice to leverage student learning within qualitative research. The main guiding question was, what does arts-based inquiry and reflective practice allow university students to discover about the nature of social research and their role as a researcher? Seventy undergraduate students enrolled in qualitative research course participated in arts-based self-study and reflective practices to help them articulate their tacit values and approaches as social researchers. Findings indicated that participating students were able to use reflective protocols to leverage self-understanding of values, influences, and abilities that helped constitute their developing identities as social researchers. Based on study findings and established reflexivity methodology, a set of general guidelines for teaching reflexive research methodology are advanced which address key elements, namely: the locus of research control, the locus of researcher exposure, the locus of researcher positioning, and the scope of researcher authority. Overall, this study is significant in revealing the power of arts-based inquiry and reflective practice in helping students explore challenging aspects of their learning and identity. These chapter findings are useful as a best practice to aid university instructors and other educational professions.

This year, through the qualitative research class and the collage inquiry exercise, I have discovered a lot about myself as a researcher. First and foremost, I have realized that you need to take a deeper look into things. For example, an image is only an image, but if you look at it in a different perspective, you may discover things you've never seen before like a color that can mean an emotion. Not only when I evaluated but also when I was creating my own collages. You want to be different and think "outside the box," and to achieve that, you need to look at the question twice in order to see and understand an aspect you've never seen before. Second of all, I realized that it takes a lot of dedication and concentration to be able to understand someone else's feeling or thought. It is perseverance that I discovered through this exercise. That's what a researcher is all about and I think I just might have those criteria inside me (student excerpt).

Over the last fifteen years, selective alternative research strategies, namely, arts-based inquiry, self-study, and reflective practice, have gradually become entrenched within educational research and professional development contexts (Bullough & Pinnegar, 2001; Leitch, 2006; Samaras, 2009; Schön, 1983). First, previous studies indicate that artistic methods can be used as a contextualizing activity when analyzing qualitative data (Butler-Kisber & Borgerson, 1997) or to visually represent study findings (Mullen, 2000). The chapter authors have found that the visual-based approaches, in particular, are well suited for instructors and students within the digital age because it aligns with the strong visual aspects the Internet and online communications. Second, self-study reflective practice are also becoming commonly employed in teacher education as a tool to explore one's identity as an educator and identify areas of personal practice that need to be refined, adapted, or abandoned (Bullough & Pinnegar, 2001; Leitch, 2006; Samaras, 2009). Third, reflexive methodology is also increasingly discussed in research teaching to better assist developing qualitative researchers to examine themselves and how they influence a research project (Alvesson & Sköldberg, 2009).

Due to the relatively short history of these alternative research strategies, research literature in many areas is lacking, particularly within undergraduate teaching research where the such approaches offer a potential tool to help students develop a better understanding of what it means to conduct social science research and the ability of creative techniques to reveal previously unknown aspects of themselves and their various role as researchers. For this reason, this study will endeavor to fill this gap by presenting the results of a study in which students were required to practice both artistic and reflective methods of inquiry to articulate their understandings of themselves, the process of conducting social research, and their philosophy as social researchers.

The central research questions this study attempts to answer are: (1) what does arts-based inquiry and reflective practice allow university students to discover about their identities? and (2) how do university students perceive the value of engaging in strategic arts-based reflective practice within undergraduate research education? Students created arts-based self-studies (collages) with accompanying personal narratives and were subsequently probed on the experiences and perceptions using a reflective protocol. The collages and narratives focused on student philosophies as social researchers. This study explores the reflections that undergraduate students produced in response to reflective protocols used following the completion of their arts-based self-studies (collages) and personal narratives. It does not examine the content of student arts-based assignments in detail since the study scope is mainly focused on the reflective responses produced by students following the completion of an arts-based assignment. As will be revealed in the chapter, reflective protocols contributed valuable insight into student perceptions on how this arts-based self study activity allowed them to understand and articulate previously hidden aspects of their identities in general and their identities as student researchers.

The study is inspired by the work of Lev Vygotsky, who explored the relationship between the development and expression of creativity, imagination, and reasoning throughout different stages of life—childhood, adolescence, and adulthood (Ayman-Nolley, 1992). In line with Vygotsky, the researchers in this chapter view creativity as an important aspect of learning and development that often gets repressed in adulthood when adult learning and life demands can force individuals to "abandon their creative efforts" (Smolucha, 1992, p. 56). This chapter is based on the assumption that creativity and self-reflection are core aspects of adult learning and university level instruction that need to be nurtured to be preserved.

BACKGROUND

Cultivating the imagination and creative learning opportunities represents an important educational aim most of us experience as students in our early years of formal school. Creative learning helps stimulate the imagination of young minds be self-reflective and open to explore new ideas and interests. As highlighted by De Bono (1970), "Creativity is a great motivator because it makes people interested in what they are doing. Creativity gives hope that there can be a worthwhile idea. Creativity gives the possibility of some sort of achievement to everyone. Creativity makes life more fun and more interesting." Creativity learning is not only important to childhood education but adult education as well. According to Vygotsky, imagination is reflected in childhood play, it is linked to cognitive processes and "is the basis of every creative action" (Lindqvist, 2003, p. 249). During childhood, children rely on elements present in reality to stand in for those in the imagination. Creativity stems from individual experience and thus children enact what they know from previous experience rather than create something new, which Vygotsky asserts is the hallmark of creative activity (Lindqvist, 2003). During adolescence, imagination becomes a "higher mental function" in that imagination is less a function of drawing on elements of reality and more a function of conceiving of new concepts by envisioning fantasies (Smolucha, 1992, p. 56). However, adolescents also develop the capacity to reason which fuses with the paths of imagination, which may prompt them to develop a critical attitude and "abandon their creative efforts" (Ayman-Nolley, 1992; Smolucha, 1992, p. 56); that is, they rely on reasoning to critique the products of their imagination to the point where creative activity is curtailed or ceased in adulthood. Neglecting creative learning in post-secondary education is a serious concern for many university instructors and teaching assistants who recognize the importance of cultivating creative skills and ability to think outside the box.

The problem of neglecting creative learning opportunities in post-secondary training is even more serious when considering that adults have a high capacity for creative thought. Following Vygotshy, adulthood constitutes the stage of life in which creative processes are most "mature" (Ayman-Nolley, 1992, p. 81). Adults' imaginative processes are richer and more complex than children's as they can draw upon additional life experience to fuel their creative thought (Ayman-Nolley, 1992; Smolucha, 1992). As a result, Vygotsky recommended exploring "different but interrelated components of creativity," including imagination, reasoning, and life experiences so as to harness, engage, and maximize one's creative potential when it is at its peak (Ayman-Nolley, 1992, p. 84). This creates a need for university level instructors to design rich learning environments that stimulate the imagination by promoting self-reflection and creative learning opportunities.

One way to leverage creative education is through the arts as argued by Elliot Eisner. Eisner (2004) states, "Artistry, therefore, can serve as a regulative ideal for education, a vision that adumbrates what really matters in schools. To conceive of students as artists who do their art in science, in the arts, or the humanities, are after all, both a daunting and a profound aspiration." In line with this, a growing body of work under the umbrella of arts-based inquiry has helped bring the creative aspects of the arts into traditional research teaching and learning. McNiff (2007) defined arts-based inquiry as the "systematic use of the artistic process, the actual making of artistic expressions in all of the different forms of the arts, as a primary way of understanding and examining experience by both researchers and the people that they involve in their studies" (McNiff, 2007, p. 29).

A number of arts-based inquiry applications (i.e., drawing, collage work, painting, drama, and music) have recently been applied within educational contexts to promote creative processes, self-reflection, and reflexivity among student researchers (Hamilton & Pinnegar, 2009; Samaras, 2009). One of the reasons for this is to promote reflexive learning among developing student researchers so they are able to understanding themselves and their influence within the research context. Advocates for reflexive methods, Alvesson and Sköldberg (2009), offer their own interpretations as to the characteristics that comprise "good," reflexive self-study. In the great divide between scientific and interpretive researchers, the authors recognize the difficulties researchers may encounter in striking a balance between methodological concerns and a focus on researcher interpretation (Alvesson & Sköldberg, 2009). However, they assert that the latter is impossible to avoid, as researcher perspective will inevitably permeate the data, thereby rendering it "theory-laden" and consistently placed within a frame of reference even prior to interpretation (Alvesson & Sköldberg, 2009, p. 6). As a result, supposed facts and research results do not point to absolute certainty or accurately reflect reality. Rather, reality is portrayed according to researcher perspectives. Thus, the authors remind researchers to pursue research with caution by remaining reflexive throughout the entire research process (Alvesson & Sköldberg, 2009).

Artistic inquiry projects have been assigned in self-study methods graduate courses, which granted students the opportunity to explore their research interests, define their role as student researchers, and gain a better understanding of themselves in relation to others via arts (Samaras, 2009). Recently, researchers have also relied on collage inquiry techniques in self-study as a means of representing their experiences as researchers, educators, and artists throughout their academic careers (Hamilton & Pinnegar, 2009). In constructing the collages, the researchers realized that collage making requires the same rigorous "strategies of analysis and interpretation" employed in previous research endeavors and constituted a unique tool to offer insight into personal processes of representation of phenomena and experience in relation to others (Hamilton & Pinnegar, 2009, p. 157).

Building on previous work, his study aligns with Vygotsky's line of thinking and answers his call for people to test the boundaries and possibilities of their creativity. That is, sharpening one's creative capacities will better equip students for future academic endeavors, artistic, alternative, conventional and scientific alike, which according to Sullivan (2006), need to "be systematic and rigorous, but also inventive so as to reveal the rich complexity of the imaginative intellect" (p. 20).

METHOD

This arts-based study focusing on reflective practice within undergraduate research classes draws on a phenomenological research design defined as a study that "describes meaning for several individuals of their lived experiences of a concept or a phenomenon" (Creswell, 2007). The study was conducted using data collected from two separate third year qualitative research methods classes that ran during the Fall 2009 and Winter 2010 sessions at the University of Ottawa. As part of their coursework, two undergraduate qualitative research classes were required to conduct a reflective self-study using collage techniques in which they explored their perceptions of their philosophies, approaches, and identities as student researchers in the social sciences. Students created arts-card collages in both hardcopy and digital formats that depicted their attitudes and beliefs in response to the question, "what is my philosophy or approach as a social researcher?" To accompany the art collages, students composed a written narrative that included information such as a description of the scene depicted in the collage, an explanation of their decision-making processes in image and material selection, and overall thoughts, feelings, and observations regarding the question posed (see Appendix 1 for an example). An integral feature of the assignment and the core focus of this chapter is on the reflective researcher diary submitted by students which responded to the following: "what did this collage inquiry exercise allow you to discover about yourself and your identity?" and "what did this collage inquiry exercise allow you to discover about yourself as a social researcher?" This chapter explores student responses generated by responses to the reflective protocols. The emerging student reflections were subsequently analyzed using basic textual analysis coding strategies prescribed by Creswell (2007).

FINDINGS

Findings indicated that all participants were able to use the reflective protocols to probe their understanding of their self-study collage inquiries. The coding yielded the five main themes that reflected commonalities among students' responses regarding the insights into their identities and research approaches gained from the exercise: (1) self-understanding, (2) nature of research and researcher role, (3) preferred approaches and techniques, (4) challenges, and (5) perceived value of arts-based inquiry and reflective practice. The themes are discussed in detail below.

The majority of reflective responses indicated that the collage inquiry exercise allowed students to achieve *a sense of self-understanding*. This is illustrated in the following student excerpts:

Looking back on the time I've spent working on my collages and looking at others, I really learned much about myself and well as our generation in general. (student excerpt)

This collage inquiry exercise allowed me to discover many things about myself as a researcher. I have discovered that I am very observant and I would rather describe anything by actual face to face interaction. (student excerpt)

The collage inquiry exercise forced me to think of broad concepts that shape my life that I had never articulated or expressed in a tangible manner before. (student excerpt)

This exercise was helpful in forcing me to look inward to express the person I am in order to understand my viewpoint on the world and how I interpret others, situations, and phenomenon. (student excerpt)

The collage activity gave me a great deal of self-awareness about how I define my self as a person. (student excerpt)

Students considered the activity as a 'process of self discovery,' an 'introspective exercise,' an 'opportunity to reflect' that permitted them to become acutely aware of goals, experiences, characteristics, desires, values, approaches, abilities, and influences which were all considered features of their identities that had previously been overlooked or forgotten. (student excerpt)

To some students, their self-understanding incorporated a concern for the future, whereby students envisioned the kind of person they wanted to be, what they wanted to achieve, and what they viewed as their future priorities. One student remarked, "I believe that I have recognized my identity and have a bright future awaiting me, along with the help and support of good friends and family. I am committed to finishing my honors degree here at the University of Ottawa and then continuing my career as a Federal Government employee" (student excerpt). Another student noted, "Delving into who I really was, and what I wanted to achieve was both frightening and exhilarating at the same time, the collages provided a new avenue for expression whereas I usually only rely on words to do the talking" (student excerpt). A third student added, "Now that I have realized that I am thinking more into my future and what my priorities will be. I want to spend more time considering my options for the future rather then getting fixated on one career path" (student excerpt). Students (re)examined the directions their lives were leading, which varied from abstract goals such as self-improvement and enlightenment to more concrete aspirations, such as continuing their careers, applying for graduate studies, getting married and starting a family. However, students also conveyed their awareness that their identities were firmly connected to the past by acknowledging the influence of past experiences, the importance of understand[ing] how they got from one situation to another, and key experiences, whether positive or negative, that helped determine how they came to be where they are today:

> The collage assignment was useful in allowing me a different way to look at myself and perhaps expose certain biases I may have. I realized how driven I am by my past experiences and how important family is to me. (student excerpt)

> Those life-changing moments are typically where there has been some clarity in our own life-purpose therefore it is crucial to understand how you got from one situation to another. (student excerpt)

Most of the time, university work is based on research and innovation so it doesn't leave room for you to self reflect and express where you came from or how you came to be, let alone artistically. (student excerpt)

Our past experiences are influences each time we write an essay or do a project and I realized how important it is to understand why we are thinking the things we think. (student excerpt)

My heritage and life experiences, and how I indentify with these, plays a huge role in my opinion formation, and consequently in how I express myself. (student excerpt).

The collages have made me reminisce about my past experiences and events that have affected my life in good or bad ways. (student excerpt)

The activity also served to illuminate many different personal characteristics that students did not know or forgot they possessed, as well as essential characteristics about their identities. Various dichotomies emerged that spoke to the diversity of personalities in the class and individuality of each student. For instance, there were those students who took pride in expressing themselves and those who were comfortable sharing certain things. There were also those who did not always communicate or express their beliefs and who considered themselves introverted and reserved. It is interesting to note that collage inquiry appeared to be an effective means of expressing aspects of identity regardless of whether students perceived themselves as introverted or extraverted. Selected dichotomies in self-understanding of personal characteristics are illustrated in terms of extroversion and introversion in Table 6.1.

Table 6.1. Selected examples of extroverted and introverted indicators.

Extroversion	Introversion
I take pride in expressing myself freely, while however approaching situations with caution and concern. I believe in doing whatever it takes to accomplishing things in life but must first conquer feelings of apprehension and anxiety. (student excerpt)	I also realized from this collage inquiry exercise that although I do not always communicate or express my beliefs, I am still very passionate about certain elements that formulate my identity. (student excerpt)
Over the semester, I have come to understand that I am a person whose identity and perception come from the people who I surround myself with, especially my family. I hold their lives highly in my heart and will always want to share my experiences and adventures with them. (student excerpt)	The collage inquiry exercise has allowed me to discover that I am a diverse, opinionated, albeit introverted individual (student excerpt).
This collage inquiry exercise has helped me to discover that I am quite an opinionated individual. (student excerpt)	Although I am a little reserved, I still have a lot to say about my identity. (student excerpt)
	I enjoyed making these art collages because I felt! could communicate personal aspects of my life to whoever the audience was. The arts—cards served as a private flat form through which! could express myself free of judgment. (student excerpt)

Beyond reflections on introverted and extroverted indicators unearthed, students also reflected on defining qualities of their personal identity such as compassion, open-mindedness, and hard working. As noted by one student, "The self-identity arts-card submission reinforced my belief that compassion is a part of who I am" (student excerpt). Another student expressed, "While my approach is still characterized by a certain amount of uncertainty, it anchored in my own values of hard work, discipline, open-mindedness and understanding" (student excerpt). In addition, a common view emerged frequently throughout students' responses in which they acknowledged that they could not isolate all of the characteristics that defined them, and conceived their identities to be "multidimensional," multifaceted" fluid and two-sided rather than fixed, stable constructs. This is reflected in the following statements:

> Finally, this exercise has shown me that I am multidimensional and that I have chosen to be everything which I am, and in that respect that I can choose to be whatever type of person I want to be. (student excerpt)

> I learned that my identity is a fluid entity, and that it changes as challenges are presented to me on a daily basis, that the things I took pride in years ago aren't the same things I take pride in now. (student excerpt)

> Doing this arts-card also reinforced my awareness that I am a textbook Pisces: two-sided identity, dreamer, secretive, intuitive, compassionate, lover of all things mystical. (student excerpt)

> I have discovered I am passionate about so many elements as well as I am defined by a multitude of elements. (student excerpt)

Self-understanding was not only conveyed by students in terms of personal attributes but also in terms of their interests, life values, and perceived influences on their developing self-identity. Material, tangible things such as family, friends, music, sports, poetry, money, and career were noted as meaningful aspects that comprised students' identities, but they also attached meaning to more abstract concepts such as time, heritage, faith, education, love, opportunity, health, peace, and adventure. In the same vein, a set of defining feature of students' identities included influences. Students indicated that elements of their surroundings contributed their identity development, and affected subsequent actions, decisions, or events, such as peer pressure, familial relationships, upbringing, travel and encounters with other cultures, formal education, and the desire for social approval. Overall, it became apparent that the collage exercise allowed students to garner a better understanding of their identities, not only in a general sense, but also in terms of specific characteristics, values, and influences, that were indicative of their identities. As is apparent below, reflections on identity were extended by students to considerations of their identity and values as social researchers.

The second theme formulated from the data, *nature of research and researcher role,* revealed a high level of student engagement in reflecting on the nature of the research process and their role within it. As indicated by one student, "Insomuch as the collages taught me about myself and my identity, it also showed me a new perspective on myself as the researcher" (student excerpt). Key aspects of social researcher positioning were apparent in students' reflections on how the collage inquiry helped them articulate their own researcher approach. For instance, some students focused on generic characteristics they perceived to be prerequisites of a successful researcher, regardless of field or orientation, such as inquisitive, organized, dedicated, and passionate: "I learned that one of the most important parts of beginning research is to organize your findings and categorize all of the elements included in the study" (student excerpt), "I realized that it takes a lot of dedication and concentration to be able to understand someone else's feeling or thought. It is perseverance that I discovered through this exercise. That's what a researcher is all about and I think I just might have those criteria inside me" (student excerpt), and "I also realized it is essential for me to be interested to almost passionate about the topic I am researching since this will lead to a rewarding learning experience for me as an individual" (student excerpt). The student focus on the general nature of social research entailed student views on what they could do as researchers based on the collage inquiry activity and the course as a whole. For example, many students indicated that they came away from the exercise better able to code data and track researcher progress through journaling as indicated in the following excerpts: "Perhaps one of the more interesting things that I discovered about myself in this exercise was that I really enjoy coding other people's collages and narratives and trying to decipher different themes and trends" (student excerpt), "The exercise was also useful in showing me a different way to research journal or even conduct research with others" (student excerpt). Some students commented on how the activity provided them a better grasp of how to "draw connections between things that are seemingly different," and sharpened abilities to "think critically and analytically" " (student excerpt). A few students alluded to the creative liberty and flexibility that this research exercise afforded them, as captured by repeated references to retaining an open mind to complexity, new ideas, possibilities, and opportunities. As noted by one student, "I think I already had certain conceptions about what type of researcher I am, or would be in this class.

It is noteworthy that data analysis also revealed a strong student focus on how their researcher-participant relationship influenced their researcher perspective and how they navigated data collection and interpretation. Many students believed that they could not be certain that the audience would understand the meaning of the message the same way they did and that there is always room for interpretation. Students realized that research was framed within the perspective of the researcher (and author), but that it was also read by the perspective held by the audience and readers. In addition, according to many students, meaning was subjective and the conclusions drawn from any research study were subject to multiple interpretations. Students also recognized how the researcher's perspective, knowingly or not, permeates the collection, analysis, interpretation, and representation phases of the research process. However, while students asserted the importance of considering the perspectives the researcher and audiences may impose on the data and its interpretation, temporarily suspending personal views and stepping outside one's role as a researcher to focus explicitly on the perspectives that participants bring to the research was also a popular sentiment. That is, students were able to reflect on the importance of perspective and researcher flexibility in order to picture to accommodate different beliefs, morals and values from participants. As noted by one student:

> I think that being compassionate toward the views of others allows me to understand their experiences through more of an insider-view because I can picture myself in

their shoes. In some cases, where warranted, this allows for a better interpretation of the participant's experiences. (student excerpt)

The third theme identified, research approaches and techniques, incorporated specific references to students' preferred approaches and techniques. The majority of student aligned themselves with a particular research tradition (qualitative or quantitative). Some students advocated the necessity of remaining objective in research by cautioning others against the input of researchers' opinion into their findings. One student stated, I need to ensure that I remain objective and learn how to control my subjectivity. I try to completely remove any other perspectives I had on the topic and start from fresh" (student excerpt). Another student aligned the role of the researcher with scientific (or objective) framework, "As a researcher I have realized that I do not agree with the input of researchers opinion into their findings. It creates bias and taints the study and therefore the results" (student excerpt). In contrast, many individuals mentioned they possessed many of the interests and qualities typical of qualitative researchers; such as exploring what people think and feel through artistic expression and stories. This type of sentiment expressed students' acknowledgement of the importance of social interaction and lived experiences as the focal point in qualitative research. One student indicated, "I am a searcher that seeks to understand interpersonal communication and I enjoy knowing what people think and feel. I found that this project really helped achieve this goal and discover how you can really understand people's feelings and opinions through art" (student excerpt). A second student added, "The type of research that I enjoy doing is where I am able to listen to people's stories in order to further understand how they came to be where they are today " (student excerpt). In some cases, it was apparent that student engagement in collage inquiry forced them to re-evaluate previously held beliefs concerning their own preferred research style and approach.

Others students aligned their research with specific research designs (i.e., phenomenology, ethnography case studies, semiotics) they had learned about through class work and reading. This is apparent in the following student excerpts:

It helped to define to me how semiotic analysis is based and essentially surrounded on subjectivity. This was very useful to me as I preciously did not comprehend how blatant and obvious signs could mean so many different things to people. (student excerpt)

I believe that my best suited type of research would be ethnography. I believe I have the ability to describe people and situations through my writing. I enjoy talking to diverse groups and would take pride in fully engulfing myself in their culture. It would give me the capability to live as something new and experience new things that are different to the culture and things I am accustomed to. (student excerpt)

Throughout the course period, learning about how a researcher portrays certain abilities and strategies used in order to evaluate case studies and research has allowed me to learn that an increase of reading and digging in for information has taken a great amount of time in order to search what you, the researcher, is trying to find. (student excerpt)

In addition, many students focused on specific research tools and frequently mentioned interviewing or observation as their tools of choice when collecting data as evident in the following "Interviewing and searching in-depth, rather than breadth, is what I find to be meaningful research" (student excerpt), "The collage exercise was helpful in many ways—Firstly, this was good practice in finding out what observations are most apparent to me and how I code them" (student excerpt). These methods allowed students to gather the richest data from which they could garner understandings of human experience, citing experiences, interpersonal communication, and why people behave the way they do as areas they sought to explore through research. Responses were also telling as to how students managed data once it was collected. To some students, analysis techniques involved searching for commonalities among participants' experiences, behaviors, attitudes, and responses as well as the researcher's observations. The overall aim of analysis was to get the underlying essence, analyze underlying meanings, explor[e] the larger themes, look at the big-picture of the data, and try and find a pattern or logical link among individual instances to better understand the phenomena under investigation or make connections to those outside of the research context. These students' statements imply that analysis partly hinges on the researcher's subjective meanings derived from the data; however, some students questioned the researcher role in the research process. As expressed by one student, "It is important however to not bias your opinion based on your observations because you could easily change their meaning into your own interpretation of their art. I need to ensure that I remain objective and learn how to control my subjectivity" (student excerpt).

The reflections indicated that students' opinions varied in terms of the degree of researcher involvement and interpretation that should be accommodated in research. In line with Alvesson and Sköldberg's (2009) reasoning, some students asserted that the researcher cannot completely separate personal biases from the data. These students concluded that interpretation was inevitable and it was "impossible to stay bias free," and thus they preferred approaches that allowed them to "bring in my personality," that was "anchored in my own values," and that created opportunities to "have personal input." Conversely, other students revealed their inclination to conduct research that was untainted by subjective interpretation. For these students, maintaining objectivity was a guiding principle throughout the research process and they expressed their intentions to "keep a distance between yourself and the content of your study," "keep my biases to the side," "control my subjectivity," and "[bracket] the researcher's beliefs." Researcher interpretation was not the only contentious issue as students' opinions were also divided as to the amount of flexibility the researcher should be allotted when conducting research.

It is important to note that students appeared to align themselves within a dichotomy of linear (objective) and non-linear (subjective) research approaches. Some gravitated towards non-linear approach[es] that permitted experimentation with different methods and accommodated new insights that contradicted expectations or changed the direction of the study, while others asserted their approaches tended to adhere to a rigid structure in which each step occurred in a sequential order. For example, one student enjoyed conducting research because of the freedom to which he or she was entitled in terms of topic selection, data collection and analysis techniques, and presentation of results. Other students suggested that research was a process of trial and error, in which one could see which approach would work most effectively for the topic at hand. Another student described the importance of flexibility as she learned to expect the unexpected despite any initial assumptions or predictions she may harbor when commencing a research project. However, other students did not share a tolerance for ambiguity and were concerned about the degree of control they felt they should exert over the research process. These students opted for a linear or scientific approach in which they work in a very systematic way, and allude to a straightforward style in order to eliminate complexity and details that could cause confusion. For these students who felt more comfortable with standards and clear expectations when conducting research, they struggled at first with an approach to inquiry where guidelines are exploratory and emergent.

A fourth thematic element that emerged during analysis concerned *challenges* students encountered in qualitative projects like the self-study collage inquiry. One of the most obvious and common difficulties students described were the constraints of employing images to represent concepts. Some students explained that it was challenging to narrow down the images and define what to depict as reflected in the following excerpts:

> I found it difficult to narrow down the images I could use for my personal collages. (student excerpt)

> It was difficult to find images along with actually defining and distinguishing exactly what I wanted to depict and represent. (student excerpt)

> I also discovered that I thought at times a picture could not fully capture what I was thinking. (student excerpt)

In addition, some students were challenged by the absence of standardized set of instructions. Without step-by-step instructions, some students felt stifled and paralyzed. Although artistic talent was not a prerequisite in this activity, students who rarely dabbled in creative visual approaches, or who preferred text as their primary vehicle of expression indicated that they felt out of place. For these students, it was a challenge to step back and consider personal aspects of their identity, about which they rarely, if ever, reflected:

> I found out that this type of free and abstract research is very difficult for me. I have a very quantitative brain, as my research project outline demonstrated to me, probably as a result of doing research for essays in other classes, which is very structured. (student excerpt)

> I find that there is something stifling about verbalizing or visually depicting what I have faith in because it's a broad question with a variety of answers. (student excerpt)

> The creative freedom to do or create anything that would constitute an art card collage was, in itself, paralyzing" (student excerpt)

It was also a challenge to step back, think about my experiences and try to put a label or an image on my values. More practice in the area of reflexivity is perhaps required on my part (student excerpt)

Not only were students able to discern their challenges in working with arts-based self-study inquiries. The activity simultaneously prompted thoughtful reflection about the lessons to be learned from such work as indicated in the following student excerpts:

As a researcher, I always thought that my style of taking in information, as well as analyzing and developing the data was intentionally objective. I presumed I read articles, journals, and books in a critical yet very objective manner. Even whilst investigating and interviewing subjects I was confidant in the fact that I was simply a tool posing barely any influence onto the topic, swaying it in no way possible. It may sound strange, but this collage inquiry made me realize that that was an impossible aim, and foolish predisposition. (student excerpt)

I found much meaning in the fact that there were no concrete rules for creating the postcards, which is a real departure from the usual university assignments with their strict academic. I think it is can be deduced that we all like to think of ourselves as unique, but we tend to feel more comfortable with standards and clear expectations, particularly when we are going to be assessment against them. This was a challenging and intriguing assignment. Thank you. (student excerpt)

Overall, while challenges were acknowledged, lessons learned through this reflective research activity appeared to overshadowed the pitfalls in helping students realize the potent role that researchers play within social research. This leads into the final main theme.

The fifth theme reveals the *perceived value of arts-based inquiry and reflective techniques* among developing student researchers. Students' general consensus suggested that arts-based, reflexive inquiry techniques are a refreshing and enjoyable activity and garnered high praise. The exercise appeared to constitute a new means of expression and a welcome departure from the usual university assignments as depicted below:

The ability to tap into the infinite possibilities pictures present was refreshing and really allowed me to learn more about myself, to dig a little bit deeper rather than staying on the surface. (student excerpt).

I found the experience enjoyable and useful. It is a nice change from essays and quizzes! (student excerpt)

After completing this assignment, even though I only contributed to it with trivial information, it felt liberating. (student excerpt).

Images were inspiring because they provoked a real meaning, and we—as human beings- were able to infer what the meaning was without any pre-education of the narratives. (student excerpt)

Responses to reflective protocols revealed that the activity had a number of value associations that students commented on. It was apparent from student reflections that the activity had a value-added component that extended beyond a typical research assignment by allowing students to explore their creative side through the arts while addressing important questions about life and meaning. The following student excerpts typify student responses concerning the value-added aspect of this particular research assignment:

> The collage exercise has allowed me to rediscover the enjoyment and fun of being artistic once again through creating each week these collages. (student excerpt)."This exercise also opened my eyes to my creative side and made me realize that I find it very easy to express myself through graphic images. (student excerpt).

> This collage inquiry exercise allowed me to see how I feel about myself and allowed me to become aware of the type of person I am. It let me to realize what my main goals in life are and showed me where I find the strength and motivation to pursue them. (student excerpt)

> This collage inquiry exercise gave me an opportunity to reflect on fundamental questions and to get to know myself better as a person. These are things that I do not prioritize in my everyday life, since I tend to feel like there never is enough time in a day. (student excerpt)

> This collage exercise allowed me to dig a little bit deeper than typical class activity have encouraged me to do in the past. (student excerpt)

There were also student views about the value of the activity beyond the realm of arts-based self-study research within research methods classes. Many students viewed the activity as an effective intervention to relieve stress, address past problems, and help individuals confront existing fears. In other words, the activity allowed students to become intimately acquainted with aspects of their identities such as defining characteristics, values, influences, and experiences that may have remained hidden or neglected. Taking the time to focus on themselves and look inward facilitated a process of self-discovery which provided students a reflective space to think about who they are and to discover answers to difficult questions that are often hard to articulate, as illustrated in the following passages:

> The second collage allowed me to channel my grief over the loss of someone I miss very much. My grandfather had a powerful influence in the person I am today and while I have written about him occasionally since he passed away it was interesting to think about what images defined him and my relationship with him. (student excerpt)

> I normally do not like talking about my emotions, however as this project was anonymous and for my eyes only, I found that writing about things or expressing myself visually helps at time to relieve stress. (student excerpt)

> I am usually a rational person but when I call a family member who I expect to be home and they are not, I begin obsessively phoning them until they answer. Doing this collage allowed me to confront that fear. (student excerpt)

In addition to value-added and intervention related value noted by students, the reflective inquiry also helped students reflect on goals they wanted to achieve in future research endeavors. One student remarked, "As a researcher, it has made me realize that I still have a lot to learn and experience in terms of theories and methods, and what

I gather now through classes like Qualitative methods, will help form a base for what I can carry with me in the future"(student excerpt). Students also reflected more broadly at their hopes and dreams in their everyday lives as noted in the following excerpt:

> Not only did it allow me an outlet in which to express my hopes and dreams, but it gave me a strong sense of just how much I do desire to attain my goals. In addition to this, the process of completing the art card served as a way for me to map out my exact goals, and reflect upon how I will achieve them. By analyzing the art cards of others, to my surprise I found that the majority of people who created them are similar to me. We are all searching for our way in life, and in a state of constant juggling of our priorities an interests. (student excerpt)

DISCUSSION

Alvesson and Sköldberg (2009) noted the tension between empirical material and researcher interpretation as a characteristic of good research because it sparks debate, encourages creativity, discourages repetition of previously held assumptions, and challenges readers and researchers alike to look at phenomena in a new way. In line with this sentiment, this chapter focused on the neglected side of this tension (researcher reflectivity) that doesn't make it into the majority of undergraduate research methods textbooks in an effort to fill a void in research teaching practice. To validate and build on the study carried out, the chapter authors use research reflectivity principles drawn from Alvesson and Sköldberg (2009) as an established conceptual base from which to gauge evidence of student reflectivity (or lack of).

Alvesson and Sköldberg (2009) posited that research, regardless of the methods used, should reflect four elements of reflective research. First, research must have an empirical orientation in which logic and rigor are apparent and practical application of theories and techniques are justified. Second, there should be acknowledgement of the research as a largely "interpretive activity" carried out by a researcher who harbors value-laden presuppositions. Third, researchers should indicate their acceptance of the impossibility of neutrality in reflective research; since all facts and data cannot evade the researcher's interpretation, research is bound to support or challenge existing social conditions or conventions based on the researcher's view, interest in the problem at hand, and research questions posed. Lastly, an author's claim of authority in accurately mirroring reality through research should raise readers' suspicions (Alvesson & Sköldberg, 2009). These elements were all found and discussed in the study findings presented above.

Juxtaposing the four elements with the real university teaching situation studied in this chapter revealed the alignment of reflexivity findings with the abovementioned elements. It also pointed to unique aspects of study findings which acknowledge the special circumstances connected to similar undergraduate research teaching contexts. This, in turn, gave rise to the following posited general guidelines for teaching reflexive research methodology. First, research does not always have a strong empirical orientation in which logic and rigor are obvious and the practical application of theories and techniques are justified. In fact, numerous qualitative research approaches are open ended and exploratory without the need (or constraint) of an a priori logic or application of theory. The myriad of qualitative approaches that exist allows for a number of possible configurations of the reflexive-empirical balance in research. This is evident in student reflections on their own preference for research that had either a stronger empirical orientation or a stronger reflectivity orientation. One student noted, "I have to remember not to let emotions overthrow the logic of an analysis and maintain an objective side to my research" (student excerpt). This diverged from reflections from students who highlighted a stronger emphasis on research reflexivity. One student stated, "One thing that I think comes through fairly strongly in all of my collages is that I would be a researcher who is extremely comfortable with shades of grey. I would not try to impose order on the world through my research where there is none" (student excerpt). Another student added, "It has allowed me to realize that I like to be interactive in my research and that I want to reach for as much diversity and representation in my data as I can" (student excerpt). Therefore *the first guideline for teaching reflexive research methodology is that researcher emphasis on logic, rigor, and practical application of justified theories and techniques is conditional upon the research balance of control between research reflexivity and the empirical orientation of research.* To this end, research instructors should design assignments (such as the one presented in this chapter) to help developing student researchers explore their individual research style and preferred orientation to research in terms of its empirical and reflexive balance. This is what we refer to as the *locus of research control* and it is core to understanding the multiple roles that researchers can occupy within the plethora of research approaches being taught within undergraduate research courses. For instance, grounded theory and case study methodology typically align with a high degree of research control and empirical orientation, whereas, approaches like autobiography and self-study align themselves with a lower degree of research control and a more reflexive research orientation.

Second, not all researchers harbor value-laden presuppositions that influence researcher interpretations in a meaningful way that needs to be scrutinized. Although, all research may be considered an interpretive activity, the researcher does not always harbor value-laden presuppositions connected to the research context. Realizing this helps to cut down on unnecessary naval gazing, save valuable research time, and highlight the core area where reflectivity is needed. Student researchers, in particular, require structured reflection that maps onto their research orientation and core research focus. As expressed by one student, "It is difficult not to place your own opinions into your analysis, especially in qualitative research studies. I feel that I would benefit from qualitative research studies in which bracketing the researcher's beliefs (i.e., phenomenological research) is part of the data analysis steps" (student excerpt). Another student indicated, "As a researcher I have realized that I do not agree with the input of researcher's opinion into their findings. It creates bias and taints the study and therefore the results" (student excerpt). It is apparent that student experiences and preferences will influence how they position themselves along a continuum ranging from impartial outsider to partial insider. Therefore, *the second guideline for teaching reflexive research methodology entails that value-laden presuppositions that may affect researcher interpretation should be acknowledged and researcher exposure to the research context should be defined.* To this end, research instructors should design assignments to help developing student researchers explore their relationship with the research content, context, and participants, along with all presuppositions that may affect the outcome of research. This is what we refer to as the *locus of researcher exposure* and it is core to student understanding of their relationship with the research context and participants to help better select appropriate research approaches and research topics that match each student's preferred comfort zone for researcher positioning. As illustrated in the above excerpts, not all students are comfortable with research where a high degree of value-laden presuppositions are interwoven into the research process. For instance, more restrictive researcher exposure from approaches such as phenomenological research and grounded theory can be pursued by some students while others can learn about research approaches with more interwoven researcher-participant positioning such as narrative inquiry and ethnography.

Third, researchers should indicate their acceptance of the impossibility of neutrality in reflective research; since all facts and data cannot evade the researcher's interpretation. However, not all research must support or challenge existing social conditions or conventions based on the researcher's view, interest in the problem at hand, and research questions posed. Although there may not be neutrality in an absolute sense, there are certainly strategies for moving towards a neutral positioning within research. As remarked by one student, "I normally stay neutral and expect the unexpected once my hypothesis is created. I try not to get distracted by any bumpy roads, and base my researches on discipline and hard work" (student excerpt). Another student noted, "I understand that in some fields of research it is necessary for the researcher to interpret those findings but I prefer more scientific approach to social research experiments" (student excerpt). Therefore, *the third guideline for* teaching reflexive research methodology *entails the acceptance of researcher non-neutrality and the selection of researcher positioning within the research context.* To this end, research instructors should design assignments to help developing student researchers explore where they position themselves within the chosen research context amidst the array of research approaches available. This is what we refer to as the *locus of researcher positioning* and it is core to student understanding of their distance from the research context and participants. Understanding this can help students better select appropriate research approaches and research topics that match each student's preferred of researcher positioning. As illustrated in the above excerpts, some students gravitate towards more distant researcher positioning as is the case with approaches such as phenomenological research and grounded theory, whereas other students prefer research approaches with more a more proximal researcher positioning such as narrative inquiry, ideological based research (feminist, gay and lesbian, Marxist, critical theory, etc.), and participatory action research.

Forth, researcher must avoid claiming authority in accurately mirroring reality through research. This is, perhaps one of the toughest lessons to impart on students— that qualitative research is not aimed at proving the truth because there is no one truth to be found. Instead, it is the adherence to research rigor, the presence detail richness, and perspective multiplicity that offers the audience a means of judging the authority and accuracy of findings for themselves. As stated by one student, "any research that is balanced and tries to include many different perspectives is probably more accurate, and also more easily accepted by a wider range of people" (student excerpt). That does not mean that any and all perspectives need to be considered, but that relevant perspectives that pertain to the object or phenomenon of inquiry are represented for the audience. A second student commented on the value of multiple perspectives learned during the class activity, "I was never one to develop multiple interpretations since I'm very mathematically oriented, but it definitely has taught me that there are different perspectives to the same object" (student excerpt). Therefore, *the fourth guideline for* teaching reflexive research methodology *entails that researchers must provide a rich level of description from all relevant angles and perspectives within the scope of the research context and approach selected to allow the reader the opportunity to judge researcher authority and the perceived accuracy of findings.* To this end, research instructors should design assignments to help developing student researchers explore the importance of multiple perspectives in contributing to researcher authority and audience approval. This is what we refer to as the *scope of research authority* and it is core to student understanding of the importance of designing studies where all relevant positions and perspectives are represented within the research communication. As illustrated in the above excerpts, some students are challenged in articulating many perspectives on topics of inquiry, whereas other student preferred a multi-perspective examination of the research topic. Understanding this allows students to choose topics and approaches that delimit the research scope and selection of approach to fit their research preferences. For instance, case studies typically demand a multi-perspective focus with many sources/perspectives to corroborate findings when writing up a research report for audiences. Other approaches, such as ideological based research (feminist, gay and lesbian, Marxist, critical theory, etc.) can focus on one guiding framework with fewer perspectives to be judged by the audience beyond.

Overall, this study is significant in that it revealed the utility of reflective protocols and arts-based inquiry in allowing students to conceptualize and articulate profound aspects of their identities and researcher values. Reflective protocols used to gather student reflections on their completed arts-based self study assignment delved into tacit aspects of their identities including personal characteristics, values, influences, and abilities that they believed comprised their identities as social researchers. Students articulated in their reflections how research, regardless of its form and content, is a product of perspective and it is imbued with meanings and interpretations produced by participants, the researcher, and audiences.

In line with Vygotskian theory, students relished the opportunity to (re)connect with their creative tendencies through alternative methods of inquiry and representation, which according to his theory of creativity, peak in adulthood. While some students noted their initial apprehension about engaging in the exercise, the vast majority of students concluded that the activity was a welcome divergence from typical university assignments designed to stimulate only logical and scientific thought processes. Their overall student view of the place and value of artistic approaches and reflective practice within research classrooms was favorable.

However, there are still scholars who vehemently oppose the classification of arts-based and self-study techniques as capable of producing legitimate knowledge, and thus certain measures and precautions must be enacted in order for arts-based, reflexive inquiry techniques to retain a foothold in the realm of "scholarly research." For example, Leavy (2009) and Barone (1995) assert that practitioners should develop standards to evaluate arts-based research separately from other forms of qualitative research so as not to "compare apples and oranges" (Leavy, 2009, p. 256). Similarly, Barone (2001) and Sullivan (2006) maintain that critics will be less inclined to question the scholarly merits of arts-based research if researchers "strive toward the highest quality in our research endeavors" (Barone, 2001, p. 27). Finley and Knowles (1995) hope arts-based research will no longer be demoted to a lesser form of qualitative research and that demonstrating the rigor, diligence, and dedication required to conduct such research will make it possible to "blur the distinctions between representations that are regarded as art and those that are regarded as research" (p. 139). To achieve this end, and based on the results generated by this study, similar arts-based self-study inquiries should be replicated and expanded in the future to properly gauge the extent to which arts-based inquiry and reflective practice should enjoy a reserved seat in the arena of scholarly research holds true across other qualitative methods classes. Hopefully, with the incorporation of similar exercises as an integral feature of qualitative methods courses, arts-based inquiry and reflective practice will continue to grow in popularity as viable tools to "expand, represent, conceptualize, and organize" interests and ideas among scholars and students alike (Samaras, 2009, p. 9).

KEYWORDS

- **Arts-based inquiry**
- **Multidimensional**
- **Self-reflective**
- **Self-study**
- **Self-understanding**

APPENDIX 1

Art-based Self-study Example

My approach as a social researcher is compassionate because I am studying life and relying on people as my source of information. I understand that everyone has individual emotions, values and perceptions that I take into account when I am conducting a research study. I believe it is important to be compassionate and empathetic because participants will be able to relate to me, identify with me and hopefully feel comfortable sharing in depth information. However, as a social researcher I must remain objective, which the images of the lenses in the collage illustrate. Remaining objective allows me to collect the most honest information and generate un-biased results.

Chapter 7

Distance Learning, Virtual Classrooms, and Teaching Pedagogy in the Internet Environment

B. Noroozi and A. K. Haghi

INTRODUCTION

The design and implementation of the virtual learning environment (VLE) or e-learning platforms have grown exponentially in the last years, spurred by the fact that neither students nor teachers are bound to a specific location and that this form of computer-based education is virtually independent of any specific hardware platforms. These systems can potentially eliminate barriers and provide: flexibility, constantly updated material, student memory retention, individualized learning, and feedback superior to the traditional classroom, thus becoming an essential accessory to support both the face-to-face classroom and distance learning. The use of these applications accumulates a great amount of information because they can record all the information about students' actions and interactions in log files and data sets. Nowadays, there has been a growing interest in analyzing this valuable information to detect possible errors, shortcomings and improvements in student performance and discover how the student's motivation affects the way he or she interacts with the software. All previous studies have used traditional supervised learning to represent the problem. However, such representation generates instances with many missing values because the information about the problem is incomplete. Each course has different types and numbers of activities and each student carries out the number of activities considered most interesting, dedicating more or less time to resolve them. In this context, the multiple instance learning (MIL) representation makes possible a more appropriate representation of available information. MIL stores the general information of each pattern by means of bag attributes and specific information about the student's work on each pattern by means of a variable number of instances. This chapter explains that 3D Virtual Reality is mature enough to be used for enhancing communication of ideas and concepts and stimulate the interest of students compared to 2D education. Web3D is a good and available platform for experimenting with creation of new tools as well as applications for tele-presence in form of 3D world's models. 3D Virtual Reality, software tools and associated Web technologies are mature enough to be used in conjunction with advanced e-learning systems. 3D-based content can enhance communication of ideas and concepts and stimulate the interest of students.

Active learning, as one of the learning strategies attempting to improve student learning outcomes, focuses on how to make students active in learning environment and to engage students in thinking and problem-solving activities. Because students generally remember only 10% in traditional passive learning environment, encouraging students to interact and be active in learning environment may increase their learning outcomes. Active learning focuses on the learning experiences and promotes students learning by doing. In addition, active learning encourages students to learn through problem solving, game, and learning activity. Moreover, activities always take place under a certain circumstance with a specific environment. Students who work or learn a specific subject in the active learning environment can improve their learning outcomes.

In recent years, teachers in a variety of educational settings have begun to make greater use of information and communication technologies in their teaching. Such technologies have the potential to make learning resources more accessible, to allow a greater degree of individualization and to make the learning process a more active one. Recent developments in desktop computer hardware and software have made feasible a new application of learning technologies, the 3D learning environment. It is argued in this chapter that such environments can situate learning to a greater degree than traditional multimedia learning resources and can facilitate individual knowledge construction through new types of learner-computer interaction.

Distance learning is a re-invented method of education, rather than a new one. Distance learning is broader than e-learning, as it covers both non-electronic (e.g., written correspondence) and technology-based delivering of learning. Technology-based learning is delivered via any technology, so it entails distance learning, too. Resource-based learning is the broadest term because any technology could be used as a resource in the learning process, where learners are active. In its early days, distance learning consisted of correspondence education, televised courses, collections of video tapes, and cassette recordings. Figure 7.1 shows a brief history of distance learning (Hamza-Lup & Stefan, 2007; Harper, Chen, & Yen, 2004).

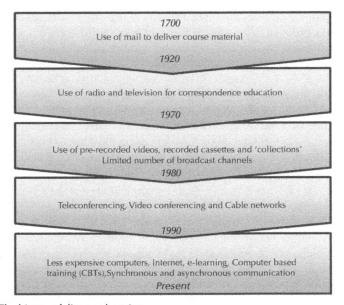

Figure 7.1. The history of distance learning.

The concept of Internet-based learning is broader than Web-based learning (see Figure 7.2). The Web is only one of the Internet services that uses a unified document format (HTML), browsers, hypertext, and unified resource locator (URL) and is based on the HTTP protocol. The Internet is the biggest network in the world that is composed of thousands of interconnected computer networks (national, regional, commercial, and organizational). It offers many services not only Web, but also e-mail, file transfer facilities, etc. Hence, learning could be organized not only on the Web basis, but also for example, as a correspondence via e-mail. Furthermore the Internet is based not only on the HTTP protocol, but on other proprietary protocols as well (Anohiina, 2005; Hamza-Lup & Stefan, 2007).

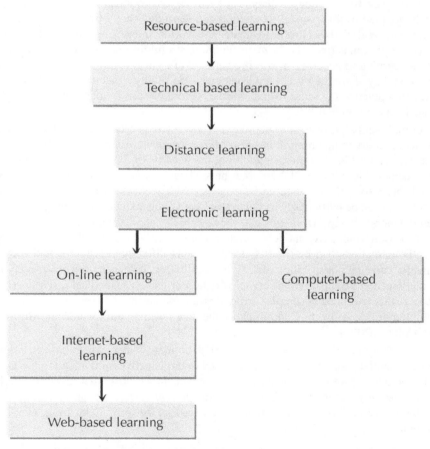

Figure 7.2. Different levels of resource-based learning.

Of particular interest is the growing number of students from developing or transitional economies studying Western university degrees. They enroll either as a foreign student at a Western university, or join an internationally accredited and qualified educational institution in their home country which collaborates with a Western university (van Raaij & Schepers, 2008).

Virtual reality and VLE have become increasingly ambiguous terms in recent years. The powerful 3D graphics hardware available in desktop computers provides an attractive opportunity for enhancing interaction. It may be possible to leverage human spatial capabilities by providing computer generated 3D scenes that better reflect the way we perceive our natural environment. 3D environments have the potential to position the learner within a meaningful context to a much greater extent than traditional interactive multimedia environments (Cockburn & Mckenzie, 2004; Dalgarno & Hedberg, 2001).

In the last two decades collaborative virtual environments (CVEs) have been largely adopted to favor social interaction and learning. They offer the possibility to simulate the real world as it is or to create new worlds. Interacting with these environments, people can actively experiment situations useful for understanding concepts as well as learning to accomplish specific tasks. CVEs support independent viewpoints for users: they share the virtual environment to do highly synchronous collaborative work, manipulating the same virtual objects. Nevertheless, it is not an easy task to create such an environment.

Virtual reality has been very popular and successful in other areas including entertainment and urban planning. It has also been extensively used within manufacturing industries and military bodies. In addition, the benefits of 3D graphics for education have been explored. Many 3D resources have already been developed in this area. 3D models are very useful to familiarize students with features of different shapes and objects, and can be particularly useful in teaching younger students. Many games have been developed using 3D images that the user must interact with in order to learn a certain lesson. Interactive models increase a user's interest and make learning more fun. 3D animations can be used to teach students different procedures and mechanisms for carrying out specific tasks. Virtual reality has also been used extensively for simulations and visualization of complex data. For example, medical disciplines use virtual reality to represent complex structures and increasingly scientists are using this technology for visualization and in particular as a teaching aid (Monahan, McArdle, & Bertolotto, 2008).

It is necessary to produce a world model, to animate avatars which have to temporally and spatially share the environment, and to implement communication facilities. Thus, the delivery of educational 3D environments based on virtual reality technologies can be very expensive, and, as a consequence, such solutions are not widely accessible to learners (De Lucia, Francese, Passero, & Tortora, 2008; Ling, Gen-Cai, Chen-Guang, & Chuen, 2003).

BACKGROUND

The term 3D environment has been chosen to focus on a particular type of virtual environment that makes use of a 3D model. Currently several trends of research are looking to progress further towards more interactive and immersive three-dimensional user interfaces based on languages like Virtual Reality Modeling Language (VRML). Many of the early 3D environment research focused on physically immersive environments, which require expensive hardware such as head-mounted displays, rather than desktop environments, which use standard computer hardware. Intelligent user interfaces are seen by many developers as the next step in the evolution of user interfaces. The single most prominent goal in the development of intelligent user interfaces is to offer a customized interface and interaction mechanism for each individual user (Dalgarno & Hedberg, 2001; Mangina & Kilbride, 2008). There are two main principles of exploring 3D models (Sperka, 2004):

1. Browser displays sequence of images (photographs or computer generated), representing panorama or the pictures of 3D object viewed from the different angles.

2. Browser calculates 2D projections of 3D object (rendering) seen from the different viewing angles.

Open systems allow everybody to create and publish 3D worlds with their own tools (VRML). Closed systems require proprietary authoring tools. Open Web3D system is the cheapest form of desktop virtual reality systems. It allows individuals to create virtual worlds and publish them on Internet for all, who have Internet connection (can be used also off-line) and simple computer with the Internet browser and any text editor. The most used are the following Web3D solutions: VRML, X3D, JAVA 3D, MACRO MEDIA SHOCK WAVE 3D, ADOBE ATMOSPHERE, VIEWPOINT, SUPER SCAPE, CULT, EON Reality etc (Sperka, 2004). As illustrated by Sperka (2004), the spectrum of Web3D applications is very broad. The list of the main existing and future application groups is presented in Table 7.1.

Table 7.1. Web 3D applications.

E-Business	Sales and marketing more effective showcase product offering to customers Shops on the web and product presentation with the virtual touch and test Real estate presentation and virtual tours Internet hotel rooms presentations Virtual panoramas of the recreation areas, ski centers, swimming pools, etc.
Education and training	3D virtual laboratories and instruments 3D e-library 3D classroom Distant education
Services	Location based services-link 3D city maps with GIS/GPS Virtual cities Hyper-markets Galleries and museums and archeological parks Airports Hospitals, etc.
Tele maintenance	Field maintenance procedures to remote locations
Data visualization	Distributed CAD/CAM systems Distributed 3D scientific and medical visualization systems e-Science
Entertainment	Internet 3D games

The use of Virtual Reality and 3D graphics for e-learning is now being further extended by the provision of entire Virtual Reality environments where learning takes place. This highlights a shift in e-learning from the conventional text-based online learning environment to a more immersive and intuitive one. Since Virtual Reality is a computer simulation of a natural environment, interaction with a 3D model is more natural than browsing through 2D WebPages looking for information. These Virtual Reality environments can support multiple users, further promoting the notion of collaborative learning where students learn together and often from each other. CLEV-R, a Collaborative Learning Environment with Virtual Reality, is a web-based multi-user 3D environment that provides a virtual reality university where students go to learn collaborate and socialize online. As with a real university, students are aware of each other within the environment and they can partake in lectures, group meetings and informal chats. Virtual Reality can bring a great deal to an e-learning experience in these ways (Monahan et al., 2008).

Communication methods provided in the system can also act as a means of social interaction for students and their peers. The system consists of a series of WebPages where prospective students can register to use CLEV-R and returning students can login to the 3D environment. Once a student provides their username and password, they are presented with a personalized webpage with information on the courses they are registered for. From this page, users can access the 3D learning environment and begin to take part in their course. The 3D environment is presented through a webpage which is split into two distinct sections. The upper section consists of the actual Virtual Reality environment while the lower section provides a graphical user interface (GUI) with tools for communication (Monahan et al., 2008).

THE VIRTUAL REALITY ENVIRONMENT

The Virtual Reality world is modeled to contain many of the features found in a traditional university. Virtual Reality is expected to enable universities and industry to make experiments in a safe and economic way, in a virtual environment rather than in the real plant where such experiments cannot be performed because of the costs and risks involved. Such experiments can provide the needed measurements and qualitative evaluations of human factors and serve as test beds of different hypothesis, scenarios and methods for reviewing the human organization and work processes, improving design quality for better safety in new or existing plants and retrospectively examining past accidents to gain hints for the future procedures (Zio, 2009).

Martins and Kellermanns (2004) believe that VLE is a web-based communications platform that allows students, without limitation of time and place, to access different learning tools, such as program information, course content, teacher assistance, discussion boards, document sharing systems, and learning resources (Martins & Kellermanns, 2004). The processes of collaboration and communication between learners and teachers are increasingly computer-mediated, such as via the Internet. From the learner's perspective, perhaps the most significant and detrimental factors to the success of a VLE are stress, association with technology use, and dissatisfaction towards the technology itself. It is suggested, conceivably, that the success of any VLE depends on the adequate skills and attitudes of learners. This proposition is evidenced by the popularity of online course delivery at postgraduate level when compared with undergraduate degree courses; as it is commonly believed that postgraduate students are mature and motivated to undertake self-study as required in most VLEs. The authors present a study with the purpose to access preparation of learners (Lee, Hong, & Ling, 2002).

Virtual Reality Environment consists of essential elements facilitating a consistent environment for learners. These elements are virtual classrooms, virtual library, meeting rooms (chat rooms), social areas and a GUI. Each room or area is equipped with tools and features to match their purpose (Bouras, Philopoulos, & Tsiatsos, 2001; Monahan et al., 2008). The lecture room or classroom is at the heart of the virtual reality environment. To create a virtual classroom, one must plan for the following tasks: advising, curriculum development, content development, articulation and credentialing, learning delivery, hardware choice, and assessment. This scenario mainly concerns the lectures given throughout the virtual environments, anticipating for virtual classrooms where appropriate teaching material could be displayed and presented by a teacher to the students. In a Virtual Reality Environment much of the structured learning takes place in a 3D classroom. The room is designed for use by tutors to address students synchronously in a live lecture. The room contains a large presentation board where the teacher can upload files in an intuitive fashion. When a tutor clicks an upload button on their virtual lectern, a webpage is presented, where they can upload a new file to the presentation board or select from a list of previously uploaded files. The presentation board can currently display PowerPoint presentations, word files and a number of image formats. The tutor can use the communication tools described above to accompany these lecture slides. Furthermore they can use audio communication to comment on the lecture slides or a web-cam broadcast to demonstrate certain points associated with the lecture. The lecture room also contains a video board, which facilitates the tutor to upload movie files to the Virtual Reality environment (Bouras et al., 2001; Harper et al., 2004; Monahan et al., 2008).

An important advantage in this system is the possibility of integration of virtual laboratories with Virtual Reality environments. The use of Virtual Reality tools does not only enhance the interaction of scientists with the experiment by offering an easy way of virtually exploring its components (human virtual reality) but also offers a solid base for rationalizing what is happening at microscopic level by providing a Virtual Reality representation of its elementary components (molecular virtual reality, MVR). Such a combination of HVR and MVR is also of invaluable help in designing innovative e-learning approaches. This scope is absolutely important in applicable sciences such as engineering, chemistry, physics and medicine (Gervasi, Riganelli, Pacifici, & Laganà, 2004).

As another element of Virtual Reality environment, Meeting Rooms have been designed to facilitate group meetings and discussions. They allow students to work together on projects and other group tasks. To this end, these rooms are equipped with similar features to those found in the lecture room. A presentation board and video board are available for students to upload their own files for others to see and discuss. When a student wishes to speak with others in the meeting room, they can use the text-chat or audio chat facilities discussed previously. One of the main differences between the meeting rooms and the lecture room is the level of restrictions which apply. Upload facilities in the lecture room are reserved for use by tutors only, thus if a student tries to upload files their actions are refused. Such restrictions do not apply in the meeting rooms on the other hand, Industry generally supports collaboration and interaction features which may include around the clock mentoring, expert led chats, peer-to-peer chats, seminars, threaded discussions, mentored exercises, discussion boards, workshops, study groups, and online meetings. Many of these activities are a simulation by technology of face-to-face interactions (Harper et al., 2004; Monahan et al., 2008).

With regard to the characteristics of communication task, media richness theory states that the purpose of communication is to reduce uncertainty and equivocality in order to promote communication efficiency. Uncertainty is associated with the lack of information. Organization creates structures such as formal information systems, task forces, and liaison roles that facilitate the flow of information to reduce the uncertainty. The role of media in uncertainty reduction is its ability to transmit the sufficient amount of correct information. Equivocality is associated with negotiating meanings for ambiguous situations. To deal with equivocality, people in an organization must find structures that enable rapid information cycles among them so that meaning can emerge (Ramasundaram, Grunwald, Mangeot, Comerford, & Bliss, 2005).

While the virtual meeting rooms and classrooms discussed above are intended for group learning and for replicating a real-life lecture scenario, a CLE with Virtual Reality also caters for individual learning with facilities for students to review and acquire the lecture notes. Obtaining the learning material is achieved in a natural and intuitive way through a virtual library. The library contains a bookcase and a number of desks. When a tutor uploads notes to the system, they are automatically represented by a book on the library shelf. Students can then enter the library and browse the catalogue of lecture notes available. When a student clicks on a book, the notes associated with that particular book are placed on the nearest free table. The user can then review the notes, flicking back and forth through them and also download them to their own computer for review at a later date. At present the library contains eight desks. If students are viewing notes at all eight desks, then the next student attempting to view a set of lecture notes is instructed to download them directly to their own computer for viewing. Personal notes taken by the student during a learning session can also be accessed via the library. In addition, a number of links to external websites are available. For example, clicking on the dictionary opens a webpage for the online version of an English dictionary, and likewise clicking on the encyclopedia opens a webpage for an online encyclopedia that students can use (Monahan et al., 2008). As a summary the specifications of a Virtual Reality Environment for e-learning can be according to Table 7.2 (Bouras et al., 2001).

Table 7.2. General concepts for collaborative virtual reality based e-learning.

Scalability	The ability to support a maximum number of simultaneous users, which could vary according to the specific settings of each virtual world.
Persistence	This is realized by distributing and synchronizing user input as well as user independent behavior in order to achieve the impression of a single shared world.
Extensibility	It should be possible to customize an existing VE, before presenting it to the users.
Openness	It should be possible to interface a Virtual Environment to external applications.
Communication	The users in a Learning Virtual Environment should be able to communicate.
Compatibility	A Web-based application implemented with international accepted standards and technologies (HTTP, HTML, VRML, and JAVA).
Support of variable content	Several forms of data should be supported and embedded in the Learning Virtual Environment.

The default settings of the 3D virtual environment smart Meeting consist of one large room, a corridor and three smaller rooms ideal for group work or two smaller rooms and a lounge with the option of listening to relaxing music and playing board games like chess. Each participant can choose between avatars with different faces and clothing. In addition, there is a personal profile where participants can provide personal information and a photo. There are limited possibilities in making the avatars show body language and face expressions. Still, they have the ability to smile, show sadness, nod or shake their heads indicating yes and no, and move their arms to wave and vote, etc. Navigating the avatars in the virtual world is done by using the mouse or the arrow keys. The avatars can walk freely inside the virtual space, sit down, move the furniture, turn the lights on and off, draw the curtains etc. It is also possible to change the default pictures and redecorate the walls in the virtual space, for example with students' work. Every participant can upload PowerPoint presentations and various other documents to display on the screen on the wall for other participants to watch. Each room is equipped with a wall screen for PowerPoint presentations and browsing the Web, watching digital video etc., and a whiteboard for drawing and writing, which is ideal for brain-storming sessions. There are also virtual laptops in each of the rooms with access to the Internet. To participate fully, each participant must wear a headset with a microphone and headphone.

Participants

The participating students are distance and on-campus students attending the same course in the field of ICT specialization, On-line Learning Environments and Distance Education.

One of the objectives of the course was to familiarize students with different learning environments. Therefore, the course provided an ideal opportunity to try out a 3D virtual environment and identify its advantages and disadvantages.

All the teachers' presentations in the course took place in smart Meeting, either in the 2D environment or the 3D environment. In addition, the teachers offered 3 scheduled hours a week for assisting students with the course assignments. In addition, many students, especially the distance students, used the virtual facilities for meetings and co-operation. One meeting was set up early in the course to introduce and learn about the possibilities of smart Meeting. The meeting began with a short presentation from the teacher and then the students were divided into smaller groups and invited to go into the smaller rooms. In each of the rooms the teacher had prepared various tasks for each of the groups that were explained in a PowerPoint document on the screen. The teacher then walked between the rooms and monitored the work and progress of the groups. Each group was asked to write a short report in Word. After the group work, everyone entered the main room and a spokesperson from each group orally presented the results

The Graphical User Interface (GUI)

The main purpose of the GUI is to host a suite of communication controls for the system. It is made up of a series of panels. The initial panel details personal user information such as their name and status. This is a messaging service where users can type a message and send it to all those connected or alternatively send it privately to individual users. The GUI also provides a panel with tools for users to broadcast their voice. Once a user connects to the communication server, they can select an area or virtual room in the 3D environment to broadcast to. For example, they can choose to broadcast to a meeting room allowing anyone in this room to hear the voice broadcast. Furthermore, users can broadcast a live stream from their web-cam to a media board in the Virtual Reality environment. The web cast is automatically displayed in the 3D environment of all users currently connected. This is actually a Face-to-face conversation tool which is considered the richest medium because it provides immediate feedback. Face-to-face also provides multiple cues via body language and tone of voice, and message content is expressed in natural language (Monahan et al., 2008; Ramasundaram et al., 2005).

A Comparison between 2D and 3D Interfaces in Engineering E-learning

It is clear that the 3D virtual environment adds something new to e-learning and teaching. It opens up a number of new possibilities, for example for group work, presentations, meetings and socialization. The results of the pilot project suggest that a 3D learning environment is a feasible possibility for parallel use with more traditional asynchronous learning environments. Until very recently the 3D was exclusively used by professionals in other fields (e.g., movie makers, web designers, etc.). It would be interesting for money traders or stock market advisers to use 3D graphics easily without spending much time learning the details of the application. A first step in this direction was taken by the 3DStock software (Hamza-Lup & Stefan, 2007; Jonasson, 2005).

Real opportunities exist for the development of novel educational and training materials, particularly for science applications where 3D visualization is critical for understanding concepts. A 3D virtual space brings advantages such as increased motivation on behalf of the student and increased efficiency in explaining difficult concepts. There are fields, such as medicine, where the Web 3D-based applications have proved their utility already (Hamza-Lup & Stefan, 2007).

The "digital classroom" provided by 2D tools does not resemble the reality of the conventional classroom, since it relies only on classical desktop windows metaphors rather than emulating real world as in 3D environments. The key feature of 3D virtual environments is the ability for lecturer/students to visualize the presence and location of other participants to the action, and, in general, to increase the presence and awareness sensation, as better described in the next section (De Lucia et al., 2008).

Dalgarno and Hedberg (2001) stated that the main specifications of a 3D environment could be as follows (Anohiina, 2005; Chen & Hsiang, 2007; Dalgarno & Hedberg, 2001):

- The environment is modeled using 3D vector geometry with objects represented in x, y and z coordinates describing their shape and position in 3D space.
- The learning process is based on some technology partly or entirely replacing a human teacher.
- The user's view of the environment is rendered dynamically according to their current position in 3D space.

The user has the ability to move freely through the environment and their view is updated as they move. Creating an online virtual classroom and providing specific learning time and space within the company may reduce the issue of inefficient use of online resources. At least some of the objects within the environment respond to user action, for example doors might open when approached and information may be displayed when an object is clicked on.

Modern multimedia technologies are matured enough allowing educators to have more options in designing the instructions. For the development of computer-based tutorials for spatial visualization skills, the comparison of 2D and 3D representations is considered necessary and informative in informing future design decision and the literature of multimedia learning (Hung, Wang, Tai, & Hung, 2007).

The potential of the Virtual Reality and 3D graphical interface as tools for delivering education is promising at all levels of educational hierarchy. However, despite all the assertions of lower opportunity cost, greater convenience, and expanded accessibility for all, it is generally accepted that the preparation of both the university faculty members and students for web-based virtual learning will need to be further enhanced. Some researchers have suggested that demands on students' time to maintain the hardware, software, and connectivity for their web-based courses are expected to be high and students might not have the maturity or skill to allocate their time productively to make progress in web-based distance education courses. Furthermore, as collaboration and communication between learners and teachers are increasingly being mediated by the computer via Internet, the likely stress and dissatisfaction about the technology itself may be the most significant and foremost detrimental factor to the success of such a VLE (Lee et al., 2002).

3D in Engineering Education

Using three dimensional graphics for more realistic and detailed representations of topics, offering more viewpoints and more inspection possibilities compared to 2D education. For example the WebTop system helps in learning about waves and optics by visually presenting various kinds of physical phenomena, may are not available in the real world, but have invaluable potential for education (Chittaro & Ranon, 2007).

In improvement of engineer education, some authors use CFD (Computational Fluid Dynamics) program to run the case and observe the physical phenomena on single-plane software rather than via internet. Frederick et al. (Hung et al., 2007) used the commercial software, Flow Lab, to create a CFD education interface for engineering course and laboratories, had been proven that is an effective and efficient tool to help students learning. Li et al. (Olivier and Liver, 2003) used the web-base to develop an on-line mass transfer course system. They also indicated that learning via internet will be the trend in the future.

The concept of thermal-hydraulics e-learning system was initiated by Hung et al. (2007). They merged existing computational fluid dynamics capability into the e-learning concept to improve traditional engineer education in fluid flow and heat transfer. In the paper, they referred to Hyper Text Markup Language (HTML) and Active server pages (ASP) where an interactive user interface as the core of the system and system structure has been preliminarily planned. The prototype of a postprocessor program was also established. But the ability of post-processor in drawing for flow field, stream lines, and isothermal contours has to be further enhanced. Liu, Hung, Hung, Pei, & Zhang (2006) quoted the concept of Hung et al. (2007) to implement and strengthen the architecture of this thermal hydraulics e-learning system. In order to establish a cross platform for e-learning, they used an HTML embedded scripting language, Hypertext Preprocessor (PHP), and integrated with MySQL database management system to manage all teaching materials (Hung et al., 2007).

In general, virtual 3-D models can be used in engineering education to stimulate processes to enhance the understanding of the concepts used in a particular process. Students could experience the 3-D structure of a specific process and the means of controlling direction and movement within a system.

CONCLUSION

This chapter has discussed the potential educational applications of 3D environments. Although the potential of 3D environments as learning resources is clear, there is still a great deal of work to be carried out before designers can be sure about where 3D environments should appropriately be used and about how best to design them. There is now a growing body of literature describing examples of educational 3D environments. However, there is a notable lack of published data from formal evaluations of their effectiveness. There is also a need for research that investigates the degree to which 3D environments do in fact assist in 3D cognitive encoding, and the degree to which such encoding aids in understanding, recall and application of knowledge. Additionally, if such environments are to become widespread, they need to be easy to use, and consequently work needs to be undertaken to derive use ability guidelines for designers.

Virtual Reality and VLE have become increasingly ambiguous terms in recent years because of essential elements facilitating a consistent environment for learners. Three-dimensional (3D) environments have the potential to position the learner within a meaningful context to a much greater extent than traditional interactive multimedia environments. The term 3D environment has been chosen to focus on a particular type of virtual environment that makes use of a 3D model. 3D models are very useful to make acquainted students with features of different shapes and objects, and can be particularly useful in teaching younger students different procedures and mechanisms for carrying out specific tasks.

The advantage of learning and collaborating in 3D is that it allows the learner to be immersed in a learning environment that is as close to the actual performance environment as the learner can get without actually being there. The 3D environment is more realistic than page turning e-learning and even more realistic than a classroom environment which is typically nothing like the actual environment in which the behavior must occur.

When immersed in a 3D environment, a person is cognitively encoding the sounds, sights and spatial relationships of the environment and is behaviorally engaged. The person becomes emotionally involved and behaves and acts as they would in the actual situation. When this happens, it allows the learner to more effectively encode the learning for future recall and provides the cues needed to apply the experience from the 3D world to actual on-the-job performance. It is learning by doing.

It also provides the opportunities for learners to be online in the same place at the same time looking and interacting with one another. This is far different than simply being logged into the same screen looking at the same slide. The 3D world provides a sense of "being there" which, again, ties to visual and mental cues which makes the recall and application of the learning that occurred in a 3D world more effective.

Virtual environments model the complexities and uncertainty of interacting with physicians in the physical world with opportunities for practice, repetition and shaping of behavior. Learning in 3D virtual worlds drives performance because it allows authentic practice and rehearsal of skills in a realistic environment with appropriate cues. The Advantages of 3D Immersive Learning include the following:

- Learner cues on visual, auditory and spatial elements of 3D environment which leads to better recall and application of learning.
- Learner rehearses on-the-job behavior in an environment as close to job environment as possible. (Realistic learning environment)
- Allows learners at a distance to be in the same place to practice behaviors (not just online at the same time).
- Learners become emotionally involved in the learning due to realism.
- Experienced learners can explore more possibilities of dialogue than in a scripted simulation.
- Sense of "being there" for the learner.

However, there are some limitations. First, using 3D animation with high resolution causes a large file size which yields to the difficult of students who use it on web. Because of the large file size, students who study the content in this active learning environment need to have high speed internet. Fortunately, the Internet technology is improved and people tend to have more high speed link. The future development of this active learning environment is to create the collaboration component which students can not only interact with the media but also interact with their friends. Therefore, the collaboration component will also help students improve their learning outcomes.

KEYWORDS

- **Collaborative virtual environments**
- **E-learning systems**
- **Graphical user interface**
- **Multiple instance learning**
- **Virtual learning environment**

Chapter 8

Student 2.0: A Look at Student Values in a Digital Age

Rocci Luppicini and Carol Myhill

INTRODUCTION

Human values shape attitudes, beliefs, and interactions in life and society. Values are shaped by the social norms and age in which people live. Student 2.0 has grown up with values emerging from their experiences within an increasing digital world where the use of information and communications technology (ICT) is a regular part in school and life. This chapter examines student's core values to find out what they value most in life. To this end, students at a large North American University created a series of artistic and textual descriptions to explore what they value most in life. The chapter examines collages and narratives from forty students to in response to the following question, "What do I value most in life?" Findings identified key values of students derived from narrative and collage data. Findings also revealed the presence of unique attributes aligned with the expression of values in visual and narrative communications. Unique value themes were expressed in visual communications that were not expressed in narrative form. Implications for educational planning and instructional design are discussed.

> The secret of a good life is to have the right loyalties and hold them in the right scale of values.
>
> —Norman Thomas (1884–1968)

> What I value most in life are my friends and family, learning, shared humanity and compassion and hope in people to make things better. I think all of these things show that I value people the most. People and the things they can do; whether that be love each other, help one another and come together to build communities.
>
> —Student Excerpt

This chapter deals with the topic of student values as within a digital age. Values play an important role in human life and society. Kluckhohn (1951) defines values as something people conceive of as desirable that influences how people act or think about events. At the individual level, personal values guide human identity development, judgment, motivation, and behavior. One's individual values help define the meaning of life and guide interactions with others and the environment. At the societal level, the prevailing shared values of people within a society can inform and guide the direction of that society and its institutions (education, healthcare, government, law, military, etc). As such, values constitute a powerful system entrenched in life and society where the quiet affirmation of individual values becomes interwoven within societal institutions constituted by individuals. Human value systems are particularly salient within higher education where the personal values of a young tech savvy generation of students are confronted with institutionally entrenched values of the educational institution they are attending. Despite the many educational advances that have resulted from the use of ICT's, educational software, and web-based applications, the same technologies have allowed unintended consequences to arise which may create barriers for higher level learning and instruction. The recent trend in college and university classroom bans on the use of laptops and mobile phones across North America attests to the institutional threat posed by technology in distracting and seducing students away from classroom learning to check their *Facebook* profile status, engage in online chats with a friend or family member, play online games, or surf the Internet. Given the substantive changes introduced by the integration of digital technology into life and society, one must also question the current state of human values from those immersed in a digital world where ICT's are entrenched life and society.

This chapter assumes that personal values are important to consider in relation to students growing up in an ever changing digital world. The question: "What do I value most in life?" is an important piece of retrospection, particularly in digital world marked by rapid progress and change that challenges the pervasiveness of social values. In a society that is consumed with material goods, virtual reality and online culture, technology mediated interaction, ICT's, outer appearances, and the here and now—often only the visible and tangible is considered. Without an interest in or influence from philosophy, religion, or perhaps the arts, one might never try to express one's own inner values.

At the risk of romanticizing the past, the chapter authors view a major departure in contemporary life and society from a century ago. Slower times allowed for deep level thought, less distraction allowed for more personal development, and closer community ties allowed for more strongly held societal values. The current digital age is characterized by rapid technological change, globalization, faster and more flexible information and communication exchange between many individuals, and greater access to information. It is also characterized by information overload, Internet distraction and addition, identity theft, information privacy concerns, online harassment, and cyberbullying. As the Norman Thomas quote says at the beginning of this article, "the secret of a good life is to have the right loyalties, and hold them in the right scale of values". The question is, what is the scale of values that drives students living in a digital world? What values are the most important to students and how can instructors use this information to inform curriculum planning and instructional design?

Scale of values is an important concept in relation to this study. Within this study, students were asked to be self-reflective through arts-based inquiry and answer the question, "What do I value most in life?" Students created personal narratives and self-study collages to explore their personal values. As is demonstrated in this chapter, students expressed their consideration of many different things as valued—from their family and friends right down to sports and pets. This study analyzes the patterns and oppositions between the visual and text based data, as well as drawing insights to inform educational planning and instructional design.

BACKGROUND

This literature review covers some of the major theories of cultural values, their definition, patterns, and how values affect action. Williams (1970) defines values in the following manner:

> cultural values represent the implicitly or explicitly shared abstract ideas about what is good, right, and desirable in a society. (as cited in Parsons & Shils, 1951)

In 1951, Parsons and Shils in *Towards a General Theory of Action* hypothesized that values are an intrinsic part of cultural systems, affecting the goals of organizations, modes of operations, attitudes, and behaviors of citizens. Parsons and Shils give the example that in a society whose members value individual ambition, the societal structures would be constructed around that ideal, meaning capitalism.

The digital world is a multimedia world where text and image dominates communication and expression. For this reason, it is important for researchers to mirror the world of experience studied. Capturing meaning from student 2.0 within this digital world can be enhanced by collection data that matches the visual and text based reality of their daily experiences. For this reason, both visual and text based expressions of personal value are drawn upon.

METHOD

The research design through which this study is analyzed is based on phenomenology and art-based inquiry techniques. Phenomenology is suited to this study since it is an attempt "to understand several individuals' common or shared experiences of a phenomenon" (Cresswell, 2007, p. 60). The study conducted was a collaborative self-study drawing on an area of art-based inquiry called collage-inquiry (Butler-Kisber, 2008). According to Butler-Kisber, the advantage of using collage is that "the ambiguity that remains present in collage provides a way of expressing the said and the unsaid, and allows for multiple avenues of interpretation and greater accessibility" (p. 268). Personal narratives were used to supplement self-study collage work to offer a more comprehensive self-study involving text and visual data. Data was collected from 40 volunteer students enrolled in a third year undergraduate research course offered at a large North American University. Data collection occurred in both the form of collage and written reflection in response to one specific question "What do I value most in life?" This classroom research reports on a self-study assignment given to students complete on their own time. Collages and narratives were digitalized by students and forwarded to the instructor to publish on a public class blog. In terms of data exclusion, only responses with both collage and text completed were analyzed. In some cases, only the collage was present, without the narrative. In others, only a narrative was available, and no collage. In cases such as these, the partial data was removed from the study.

An inductive coding technique was used, as described by Strauss and Corbin (1990). The data was read through once for understanding, then again with the purpose of generating headings for themes. Simultaneously, significant quotes were highlighted which appeared particularly enlightening. This process is termed horizonalization by

Moustakas (1994). These significant quotes helped form the basis of repeating themes, and develop "clusters of meaning" (Cresswell, 2007, p. 61). The codes were created using the same language as the participants. This was to help maintain the integrity of the data, and allow participants voices to be heard at all stages of the research process. This same coding process occurred for both the narratives and collages.

FINDINGS

Student 2.0 Key Values in Life

The analysis of narratives and collages revealed a range of values giving rise to the following shared themes: Family/Friendship, Education, Nature, Health, Food, Compassion/Equality, Self-Made, Art/Music, Money/Materialism, Honesty/Respect, Sports, Happiness, Religion, Life Itself, Travel, Culture, Pets, Media/Television Show, Technology, Job, Diversity, Positivity, Community, Death, and "Words". Figure 8.1 provides a breakdown of the key values found in the narratives and collage work:

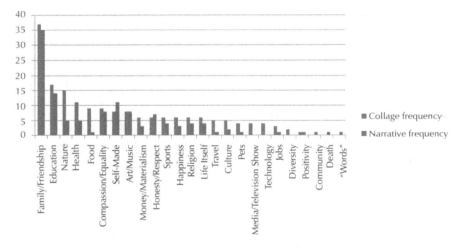

Figure 8.1. Frequency of key student values in collage and narratives.

It is interesting to note that the majority of students did not answer the question "What do I value most in life?" in a narrow sense. Most students did not interpret this question as needing a single answer and shared many things they value. While some participants defined value as what they most cherished, or thought was most important in life, other participants defined values in a traditional sociological sense as ideals.

Figure 8.1 indicates the patterns and oppositions found between the collages and the narratives. One obvious pattern is the high level of importance placed on family and friendship—this remained steady through collage or narrative. The family unit seems to have taken on a certain type of ideal status. This could remain consistent over participants since most people from all different walks of life revere their family unit. One student reasoned, "What I value most in my life is first and foremost my family. Your family will be there for you no matter what happens in your life" (student excerpts). The high value placed on family was the most common theme in the value statements and collage work. Students appeared to place a high importance on their various family relationships as seen in the following student narrative and accompanying colllage:

There are many things in life that people think they should value: a house, car, money, social status; the materialistic things that won't be there in the end. I may have valued those items when I was younger and didn't know better, but my past experiences have changed my perspective on life. Though I may not choose to associate with most of my relatives, my immediate family is what matters most. I'm always

looking out for my brother and doing whatever it takes to be there for my parents. Their happiness is what matters to me so I'd rather put them first before myself; I want their support for years to come. (student excerpt)

The second most predominant theme in both collage work and narratives was education. Students place a high value on their education and what it represented to their life and future possibilities. On student indicated, "I am a firm believer that education is what makes this world go around" (student excerpt). A second student, valued education through specific reference to where it is derived and what it provides, "I value books because knowledge is the key to escaping ignorance" (student excerpt). This theme of education was captured in collage and narrative form as depicted below:

The answer depicted is my literacy. I am in love with writing, and my major place of inspiration is the bay photographed in the piece. My book collection is featured, as well as my opinions about writing. I think reading constantly is very important for writers, hence the books. Writing is also a fight (and a battlefield), a joy, and takes both whimsy and skill.

Another predominant theme in both collage work and narratives was self-made. Students place a high value on their autonomy and what it represented to their life.

On student indicated that: "Although all aspects of life have their ups and downs, as long as at the end of the day I am able to reflect on my life and believe that I am living it the way I want to, surrounding myself with the people I love as well as feeling fulfilled and accomplished as a person, my life will always have amazing value" (student excerpt). Placing value in one's own self-determination was a moderate theme among students as expressed in collage and narrative forms as is illustrated below:

I value a lot of things in life; however, my strength and independence are what I value the most. The reason why my collage contains pictures of various iconic women, such as Oprah Winfrey and Hilary Clinton, is because they each signify the power and independence that I so value. If it were not for these women, the opportunities that I have been given and have seized would not exist.  Although it is not visible, the background of my collage is a powerful quote which I believe helps to sum up the kind of strength and independence that I am talking about: "A strong woman knows she has enough strength for the journey, but a woman of strength knows it is in the journey where she will become strong." My mother always taught me to rely on oneself to achieve success and happiness and I think that was the most important life lesson I have ever learned yet.

Variations in Visual and Text Based Value Expression

Findings revealed a number of variations in visual and text based value expression from students. One value that was strongly expressed through collage but less through the narrative was nature. People often had images of nature scenes within their collage, but never expressly mentioned nature as something that they valued. Elements of nature are evident in collage work as depicted below:

The collage above seems to show that nature is valued, yet their narrative does not mention nature. Another interesting opposition found between the collage and narrative is on the subject of food. Narratives rarely mentioned food, but many collages had pictures of food items. It is interesting to note that even in a North American society with a relatively high standard of living that takes food for granted, the value of food is known to us tacitly as core, and emerges through the collage as depicted below:

Unique Aspects of Visual Communication of Life Values

Beyond apparently similarities and differences between narrative and visual expression of values, it is interesting to note the structure of the collages as a source of meaning. Many of the collages have the structure of billboards littering New York Times Square. Completely separate ideas sit alongside each other with no interaction:

This collage work stands in sharp contrast to the collage work by Butler-Kisber (2008). Butler-Kisber's collages captured a holistic feeling, a surreal vision of a scene, odd and nonsensical, but with ideas interacting within it. The visual difference can partially be traced to how each collage was created. Butler-Kisber's examples have individual pictures that are cut around the edges, whereas the students in this collage work, for the most part, left square borders around each image. The square borders around each image, as well as students repeated use of cut out words, is one indictor of student struggle with visual expression. This interpretation is validated by Butler-Kisber's (2008) in quoting Bachofen (1967): "the symbol evokes imitations, language can only explain. The symbol touches all chords of the human heart at once, language is always forced to keep one thought at a time" (p. 49). The effect of square borders around pictures means no interaction between images, and without a greater scene in the collage, the collage is still "forced to keep one thought at a time". As Butler-Kisby suggests of collage, "novel juxtapositions and/or connections, and gaps or spaces, can reveal both the intended and the unintended" (p. 268). In regards to the above-mentioned structural aspects of the students collages, the gaps between images and lack of a holistic picture in many collages reveals an apparent segmentation of self and a struggle to self-reflect and represent personal values within visual communications. This is not surprising, given the overemphasis on text based data in the majority of research methods textbooks available within the social sciences and humanities. Like the analogy of the person who lost his keys in the buses but searches under the lighted lamppost, the challenges of subjectivity and ambiguity involved in the interpretation of visual data has unfortunately left it on the periphery of research teaching (as an alternative form of representation) rather than at the centre of research teaching where meaning is expressed and examined. This harkens back to Eisner (1990) and discussion of the potential virtues and limitations associated with alternative forms of representation.

Despite the student struggle in letting images articulate a holistic and connected picture of their value positions, the frequency chart reveals that in fact more thematic interpretations were pulled from the collages than from the narrative. Two reasons for this are media richness and the convenience of image in the global culture of a digital world. First, images are rich in meaning. It is perhaps apt here to refer to the adage, "a picture is worth a thousand words". In other words, visual communication is recognized as meaning rich and a powerful means of capturing complex ideas such as an individuals life values which require meta-cognitive awareness, self-reflection, and reflexivity. Second, the digital world is inundated with popular and convenient imagery shared by many around the globe. In mediated communication in the past, there were strong community ties and values were closely entrenched within societal structures within one's geographic area and community. This contrasts with life and society in the digital world where values and meaning are mediated at a global level through mass media and ICT's. A case in point is in regards to the actual images chosen. This apple, pictured below, was found within the collages several times. A search on Google images of the term "health" reveals that this image is the second result out of 827,000,000 total "health" images:

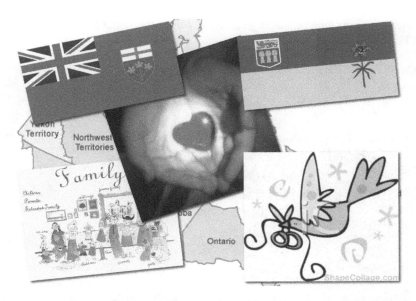

The fact that quite a few images were repeated throughout student collages points towards the role of globalization and a culture of convenience. Instead of students searching through multiple pages of results, many students took quite literally the first image they saw. Both points reinforce the idea that visual imagery is relied on by students for convenience and the ease of expression. This accounts for a higher saturation of value themes in the collage data compared to the narratives. It also reinforces that notion that student 2.0 struggles with self-reflection and deep level thought processes within a digital world of information overload and rapid change.

DISCUSSION

Study Significance and Limitations

This study is significant because it advances knowledge on student values emerging from their experiences within an increasing digital world where the use of ICT's are a regular part in school and life. Findings identified key values of students derived from narrative and collage data, along with unique attributes aligned with the expression of values in visual and narrative communications. Despite student's challenges with self-reflection and the articulation of life values, this chapter points towards student 2.0 values and habits within the digital world. The unintentional message that participants conveyed through their collage creation indicated one value, never mentioned, but highly important in the digital world—convenience. The speed, ease, and convenience of Google provided students with images and meaning with which to express their values. The fact that this is where students are going to get their information about their personal values is particularly revealing about the power and influence of the digital world.

In terms of study limitation, some collages suffered from a misunderstanding between the student and the nature of the project. Sometimes an overabundance of text was used in lieu of images, thereby removing part of the visual power inherent to collage. For instance:

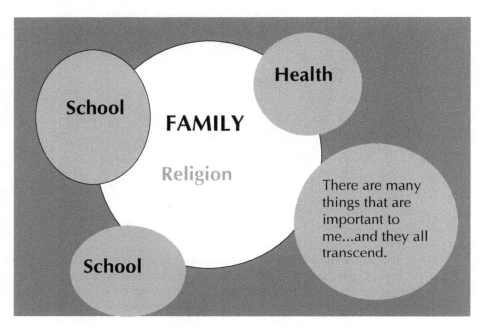

Also, some collages suffered from an overly literal matching of text to image, rather than using collage towards an end that allowed for new meanings to arise. Some students used their narrative description and literally, cut out pictures of what they valued, and paste them together like a list on a bulletin board in some cases. Butler-Kisber (2008) advises against creating a collage so literally, and instead advises, "it is useful to work conceptually rather than literally choosing images that stand metaphorically for an idea (James, 2000), and to experiment with size, color, texture, overlap, and spaces to portray the nuances of the focus" (p. 270). All of these aforementioned criticisms are potential limitations of this arts based study, yet they point towards interesting conclusions, discussed in the findings.

Implications for Educational Planning and Instructional Design

One of the most interesting elements that emerged in this study is the discovery that many university level students have not had adequate opportunity for self-reflection in their educational training. One student candidly wrote, "As the collages get more broad, I feel myself becoming more resilient [resistant] to label myself through pictures or words. This one was especially hard as I had lots of trouble trying to grasp what exactly a "value" in life was" (student excerpt). If one of the main objectives in higher education to produce responsible and autonomous adults who are self-directed learners capable of higher level thought processes and critical skills, there appears to be need for curriculum changes that help to promote reflexivity and self-reflection among students. Given, the high level of student distraction and endless amount of information available online, student 2.0 is in even greater need for instruction that promotes self-awareness to help mitigate against an entrenched multi-tasking orientation to experience within the rapidly changing digital world.

KEYWORDS

- **Information and communications technology**
- **Narrative**
- **North American society**
- **Phenomenology**
- **Web-based applications**

Chapter 9

Interweaving Education, Technology, and Life: University Student Perspectives on Icts and Personal Relationships

Rocci Luppicini and Janine Moir

INTRODUCTION

The personal use of information and communication technologies (ICTs) at work and school are blurring the boundaries of work and life within this digital age. This chapter examines student perspectives on ICT use and personal relationships. Undergraduate students at a large North American University created a series of artistic and textual descriptions to explore their personal views on ICTs and their relationships. The chapter examines responses from 63 participants to the question, "How does the use of ICTs affect my approach to relationships?" The data analysis revealed that students found ICTs to be incredibly useful in helping them, stay in touch with family and friends across large geographical distances, as well as communicating with professors and other students in their class. The chapter highlights the paradoxical nature of technology within the context of personal relationships. A number of positive and negative aspects of ICT influence on relationships are identified. The chapter provides insight into the powerful influence of ICTs on student relationships and online behavior both within and beyond the university walls.

> My closest relationships are maintained with the help of ICTs. My sister lives in Korea so I rely on Skype and MSN messenger to keep us in contact.
>
> —Student Excerpt

ICTs have penetrated our culture in everything we do, from business, to school, and personal relationships. Contemporary society uses ICTs to varying levels in everyday life, and everyone is affected by them, whether they see themselves as active participants or not. The use and dissemination of ICTs can be described as nothing short of astounding. In 2007, worldwide mobile telephone subscriptions reached 3.3 billion, "a number equivalent to half the global population," (Massimini & Peterson, 2009, p. 1). In addition, in 2007, news reports estimated that there were over 150 million active accounts on *MySpace*, and *Facebook* reported over 50 million user accounts (Scalet, 2007). Based on these astronomical numbers, ICT fluency is needed to function effectively in today's ICT-centric society. In order to use ICTs effectively, one must develop strong ICT user skills. Statistics Canada commissioned a study to see where Canada fits in to this skill-set on a worldwide stage, and how we use these ICT skills in everyday life; in work, school, and our personal lives. This resulted in the

"Connectedness Series, a set of analytical studies and research reports focused on connectedness," (StatsCan, 2005). The *Connectedness Series* covers the technological use and pervasiveness of "telecommunications, broadcasting, computer services, and Internet Service Providers," and their use across economic and social enterprises including education (StatsCan, 2005). Because of such interest in ICT integration into work and everyday life, ICT use is becoming increasingly more important to being successful in every day professional, academic, and personal interactions. This is especially true when it comes to training new university students for future careers. At the university level in this increasingly digital world, a student cannot function successfully without at least some mastery of ICTs.

There have been many research studies conducted over the past 15 years that focus on ICT use in educational and society. One particularly interesting area of research deals with how ICTs affect the negotiation of social space and how people approach relationships. It is important to understand just how inherent ICT use is in the basic functioning of North American society. In Gerald Sussman's (1999) article, titled "Urban Congregations of Capital and Communications: Redesigning Social and Spatial Boundaries," he explains the nature of ICT function and society as such;

> The infrastructure of information and communication technology (ICT) has co-evolved with industry and transportation as a central property of the metropolis, bringing new symbolic landscapes and cultural meanings to urban life. Local, national, and increasingly globalized information and communication networks have broadly inflated a range of social and spatial boundaries, that is, the reach of active social connectivity, yet they remain unmistakably centered in the metropolitan core. (p. 1)

Sussman, and others are concerned with how ICT use has affected the human use of space, and how that, in return, has affected human relationships. The point being made is that ICTs have effectively removed the obstacle of space and time in personal communication, thus blurring the boundaries of work and life, and removing the clear separation of the personal from the professional. This is partly due to the ease of public access to a multimedia platform of current ICT, which provides the public with advanced tools for managing personal and social relationships at a distance. ICTs allow individuals to exchange text, images, and film via asynchronous communication and carry on conversations in real time with full visuals, text, and voice making individuals. The authors in this chapter were drawn to the potential of ICT technology to act as a conduit for meaningful exchange between people in personal relationships. A number of interesting questions arose: How does ICT use affect how students navigate their relationships with peers, family members, and instructors? What are students experiences and attitudes concerning the use of ICTs to maintain personal relationships at school and work? Do ICTs encourage the development of personal relationships or impede them? Do students change their communication habits or styles based on their usage of information communication technologies? This chapter explores student perspectives on ICT use within relationships. It is important to examine this social phenomenon in order to better understand the experiences and attitudes of undergraduate students today and how the everyday use of ICTs affects their approach to relationships and scholastic endeavors. As stated by Pauleen (2001), "Research shows that the development of personal relationships between virtual team members is an important factor in effective working relationships." This information will be useful in identifying new opportunities to explore the implementation of digital education platforms within undergraduate university education. The following central question guides this study: How does the use of ICTs affect undergraduate students' approach to relationships?

BACKGROUND

More and more, undergraduate student are connecting with those around them and fostering personal relationships using the available technological tools, particularly through social networking sites (SNSs) and ICTs. After all, who does not have a *Facebook* or *MySpace* profile? In this digital age, it seems that regardless of whether relationships are born from family and friends, peer groups at school or university, or professional contacts and colleagues, ICTs appear to be somewhere in the mix of interactions. Recent research focuses on personal relationships in connection with SNSs and the implications this has within an academic setting where there is a blurring of lines between academic and personal life. For instance, Selwyn (2009) discusses the potential use of SNSs in the development of WBT and educational resources, through an in-depth research study based on the activities of 909 undergraduate students. Selwyn is particularly interested in the potential positive uses of SNSs to increase engagement in the academic community of students, their instructors, and the course material. He argues that many educational professionals believe that there is room for both traditional and social network engaged learning in the classroom, with SNSs providing a basis for digital education, by encouraging collaborative learning, and increasing student's academic motivations. Selwyn elaborates:

> The prominence of SNSs in the lives of learners of all ages has prompted great enthusiasm amongst some educators. It has been claimed, for example, that social networking applications share many of the desirable qualities of good 'official' education technologies—permitting peer feedback and matching the social contexts of learning such as the school, university or local community. (2009, p. 158)

Selwyn examined several hundred different *Facebook* wall posts to gain better insight into the way students appear to be naturally using SNSs. An example of one wall post examines the response of two students discussing their upcoming class essay; 12:50 pm on January 14, 2007:

> Coincidence—was just attempting to write that bit myself!! Not really too sure what to put. In my opinion it doesn't say that much about social science in general. Just a particular philosophical perspective used within it—that postmodernism is pretty much meaning—less! Beyond that I don't think there's too much more to say—except about getting some—one who works in the field to check any topics you're not sure about! But I reckon that's an implication to the whole of academia—not just social science! I just can't believe he didn't do a lecture on this—It definitely needed one!

By studying these *Facebook* wall messages, Selwyn was able to examine educationally based conversations taking place between students without prompting from their instructors or institutions. He found that students have a natural affinity to connect and identify with each other through SNSs, sometimes discussing learning situations in class, and finding that students would use to *Facebook* to "go over their experiences of recently finished lectures," (Selwyn, 2009, p. 161). Selwyn concluded that the use of *Facebook* has become an "important site for the informal, cultural learning of 'being' a student, with online interactions and experiences allowing roles to be learnt, values understood and identities shaped," (2009, p. 171). This shows that the value in SNSs should not be overlooked when evaluating important academic experiences and the students' "meaning-making activities," as SNSs have the potential to create a lasting positive influence on students' educational experience (2009, p. 171). By integrating an academic learning community that involves students, their peers, and their instructors, universities may be able to use ICTs to leverage student engagement, and student/instructor participation in and out of the classroom.

This chapter describes a study carried out within an undergraduate research class within a large North American University. Students created a class blog dedicated to the exchange of personal narratives and art collages in response to the following self-study question: "How has the use of ICTs (e-mail, SNS, mobile, Internet) affected how I approach relationships?" With this question in mind, this chapter will show how ICTs play a powerful role within undergraduate student life, relating directly to their interactions with friends, family, and peers, in and out of an academic setting.

METHOD

This study draws on phenomenology and collage inquiry techniques to explore collaborate self-studies of ICT influences on their personal relationships. Phenomenological research aims to study the meanings of human with an emphasis on "the study of lived experiences, on how we read, enact, and understand our life involvements" (Eckartsberg, 1986, p. 1). Collage inquiry is a type of art-based inquiry which is broadly defined by McNiff (1998) as the "use of the arts as objects of inquiry as well as modes of investigation." Weber (2007) adds, "by its very nature, artistic self-expression taps into and reveals aspects of the self and puts us in closer touch with how we really feel and look and act, leading, potentially, to a deepening of the self-study." This study draws on collage inquiry described by Butler-Kisber (2007) as an artful approach to communication with the potential to uncover communicative talents (that might otherwise be overlooked) and transcend the linear world of text to allow for "meaning to travel in ways that words cannot," thus uncovering both "conscious and unconscious ideas" that are often under represented in text (p. 269). To this end, the chapter draws on collage inquiry and personal narratives to explore how ICTs have affected the approach of undergraduate students to personal relationships. In terms of study participants, 63 volunteer participants in this study were students enrolled in an undergraduate qualitative research class at a large North American University. The average age of participants was 22 years and all students in this course regularly used ICTs in their coursework and were considered a well-suited population to study the effects of ICTs. Data was collected from this self-study using the following steps:

First, students created collages and descriptive narratives in response to the following question: How has the use of ICTs (e-mail, SNS, mobile, Internet) affected how I approach relationships? Second, participants then e-mailed their collages and narratives to the instructors which were posted on a public self-study art blog. Third, collages and the narratives were analyzed. Common themes and significant details were uncovered illustrating how the university students believed ICTs affected their relationships. Clusters of meaning were then developed with the data, and tables were created displaying the significant statements as developed themes about the common experience of participants in the study. These themes were further explored to get at the essence of the phenomenon of ICT influence on student relationships.

FINDINGS

Personal Narratives

Overall, these numbers show an overall positive perception of the influence ICTs have on personal relationships (see Figure 9.1).

The data shows that the majority of students find ICTs to have a positive influence on their personal relationships in providing an increase in the ease and availability of keeping in touch with loved ones. The number one use of ICTs for personal relationships stated was is to stay in touch or stay "connected" with friends, family, and loved ones while they are at university. ICTs have an overall positive influence on these relationships, and they indicate that ICTs facilitate closer relationships, are cost-effective and easy to use. There were also a number of negative and mixed perceptions of the influence ICTs have on personal relationships (See Appendices for a complete breakdown of themes). A summary table of key themes derived from student narratives appears in Table 9.1:

Table 9.1. Summary of key themes derived from student narratives.

Key Theme	Explanation
ICTs help individuals maintain connectivity between family and friends in ways not previously possible.	Within the majority of narratives analyzed, participants expressed how ICTs have affected their relationships in a positive way by helping them stay connected with family and friends anywhere in the world. Because of ICTs people can now have intimate relationships no matter how far away they are from each other because of the Internet and its applications (e.g., *Skype*, MSN, *Facebook*, etc.)
ICTs make maintaining relationships easier and more efficient.	The majority of participants mentioned in their narratives the use of ICTs have helped maintaining relationships a lot easier due to ease of access, communication flexibility, and speed.
Use of ICTs has some negative influences on relationships.	A moderate number of participants noted how ICTs take away from face-to-face interaction and lead to information overload, miscommunication, superficiality, expectations of constant connectivity, and stress.
Use of ICTs is paradoxical and has mixed influence on relationships.	A small but substantial number participants indicated mixed views on ICT which highlighted both positive and negative aspects of ICT influence.

As indicated in Table 9.1, students expressed overall positive interactions with ICTs, as they found themselves maintaining and increasing the quality of their relationships using ICTs. As one male undergraduate explains:

The use of ICTs has a positive affect on my approach to relationships. I am able to communicate with my friends and relatives across the boundaries of time and distance, which strengthens our bonds. We are constantly sharing information through a variety of communication forms such as talking on the phone, texting or chatting on SNSs, which helps us learn more about each other and ensures we stay connected. The circles of people in the collage represent this sense of connection and the bridge displays 'bridging any gaps' to bring me closer to others. (student excerpt)

Another student (female) indicates that:

The use of ICTs has greatly influenced the way I approach relationships. Since ICTs have been a part of every day life for the majority of my lifetime, it is hard to comprehend not having access to these communication tools. In my personal life, social networks like *Facebook* have allowed me to interact and stay close with people I otherwise might have difficulty interacting with. (student excerpt)

Some negative themes did emerge, however, which spoke to student frustrations with the lack of intimacy involved with ICT use and personal relationships. As one female student states:

ICTs create a hindrance within interpersonal communication and the Internet is quickly becoming a threat to traditional communication. In terms of my own opinion, ICTs have barricaded intimacy between relationships and have broken down the walls of openness. In other words, although ICT users are more outspoken, honest, and open to express their feelings, ICTs will never and should never be considered an intimate method of communication. (student excerpt)

Perhaps the most interesting findings come from mixed student views of ICT influence on their relationships which seemed to indicate a sense of conflict or duality in how students viewed their use of ICTs:

There are times where I am with friends, BUT we are all calling or texting or e-mailing someone else. In this sense, we are all together, yet our thoughts are on those far away from us. (student excerpt)

ICTs have permitted for relationships to now expand the globe over, YET still leave an uncertainty in whom the other truly is. (student excerpt).

ICTs have the power of connecting you with people who are not of close proximity to you BUT they can also lead to information overload which can negatively impact relationships. (student excerpt)

Openness and instability of ICTs have assisted my relationships BUT have also threatened them. (student excerpt)

Collage Findings

The visual data analysis revealed the occurrence of key markers in the students' artistic collages which provides insight into the influence of ICTs on personal relationships. Data analysis revealed the occurrence of several codified images. They are the depictions of children, family, love, ICT devices, global imagery, time, space (distance), dialog (words), and SNSs. These were the most obvious and frequent images shown throughout the dataset, and every single collage had a minimum of one codified image, with most of the collages having several of each code (See Table 9.2).

Table 9.2.

Key Theme	Explanation
Images of people appearing connected and happy.	The majority of the collages depicted images of people interacting who appeared close together, often happy and smiling with images of communications technologies in the same collage making them appear as though ICTs affecting their relationships in a positive manner.
Pictures of ICTs.	Images of cell phones, computers, and SNS were a predominant theme in most collages analyzed.
Images implying distance between people.	Images that showed that there was distance between people in the world. Some images reflected negatively on this distance with visual divisions or lines, while others were positive images showing people using ICTs.

Many of the collages portrayed images of people appearing connected and happy with positive body language and expression Below is one example of this:

The caption "You're still with me even though you're far away" underline images of hands connecting and people embracing. The image of the telephone, pen, and cell highlight the presence of technology.

A second predominant theme revolved around technology at the core of the collage. This is evident in the collage below:

This collage provides labels "journalism," "connect," "freedom," "talking," "e-mail," "close," "open," and "Internet" juxtaposed among images of media with a small captions of a closed hand with a note titled, "shy." The images of technology, people, and wires highlight the capacity of technology to connect people to information and overcome human tendencies toward shyness which normally would prevent such connections.

Finally, were also a number of collages which depicted divisions and distance between people where technology was seen as a means of separating people rather than bringing them together. This separation is illustrated in the following collage:

In the above collage, the dark jagged line division between the online and offline realm with corresponding labels, "unoriginal" and "original" mark a perceived distinction between online and offline. This duality is further supported by the formal figure (guard) image and sign (open). The narrative furthers this by highlight the "impersonal" nature of technology mediated communication compared to "personal" face-to-face communications. Overall the collage themes paralleled the themes extracted from the personal narratives. First, there the majority of images were positive and technology was portrayed in a positive relation to people within the collages. Second, technology was portrayed as an important part of such connections between people, making relationships possible that would not be possible without such technology. Third, there was visual evidence of a negative side of technology influence on relationships that is present in much of the collage work. This appeared to illustrate a contrast or duality in how students view the role of technology in personal relationships. This duality is discussed further in the next section as the *paradox of technology mediated relationships*.

PARADOX OF TECHNOLOGY MEDIATED RELATIONSHIPS

There is evidence in this study of the paradoxical nature of technology mediated relationships. On the one hand, most of the personal narratives emphasize how ICTs are a great way to maintain relationships all over the world, and are a much faster and convenient way to communicate. As stated by one student, "ICT opens the door to many opportunities we may not have without it such as long-distance relationships" (student excerpt). Another student adds, "ICTs make communicating with friends globally much faster and easier and as a result many long-distance relationships have been maintained" (student excerpt). Gershon (2008) notes that "these technologies allow people who might never previously have met to meet, or enable people to sustain relationships over long distances" (p. 13). On the other hand, the many students also realized that relying on technology to develop and maintain relationships can lead to impersonal relations where misunderstandings arise. One student indicated that "communication through these mediums does not come close to the richness of face-to-face interaction" (student excerpt) while another believed that "there can be confusion or misunderstandings through ICTs because of the no face-to-face interaction" (student excerpt). A third student focused on technology, noting that ICTs make "conversation less personal and less intimate" to maintain relationships (student excerpt). Other students alluded to the challenge of overindulging in online communication at the expense of face-to-face relationships. This links to a study by Milani, Osualdella, & Di Blasio (2009), which states that "research on Internet addiction has demonstrated how the tendency to use the Internet in an uncontrolled way is connected to certain personal characteristics, like the preference for solitary activities and low social openness" (p. 681).

This highlights how engagement in relationships through communication technologies is a double-edged sword that can have positive and negative influences on students. This duality is evident when looking at collage and accompanying narrative from one of the student participants:

Collage	Personal Narrative
	In some ways ICTs bring me closer to people, however I mainly keep strong connections with people that I see face to face. I am constantly invited to events on *Facebook* and make vague plans to reconnect with lost friends. Although this fills the foreground of social platforms, penciling people in on 'sticky notes' isn't always functional. It's often a way to deflect a sociably acceptable ritual of saying yes to every social concept. In reality SNSs provide me with a series of weak ties to a large body of people, which is fun, yet distant. (student excerpt)
	The image of two hands shaking through computer screens represents the globalization aspect of communications and how there is no separation between people anymore, BUT there remains a spiritual gap. (student excerpt)
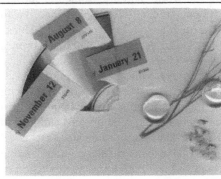	They can bring people closer together although they are far apart. I have found that my relationships seem much more emotionless (reflected by the stems with no flowers) with these technologies and the use of text messages and Internet never truly reflect how the relationship is in person. (student excerpt)

This conundrum, as revealed within collage and narrative analysis is what this chapter describes as the paradox of technology mediated relationships that students are faced with when relying on ICTs for developing and maintaining relationships. This has ramifications for education and online learning design which are elaborated on below.

DISCUSSION

Findings indicated a generally positive outlook on ICT influences on personal relationships among undergraduate students. The majority of students in this study indicated that they used ICTs on a regular basis to stay connected with their friends and family while time or distance separated them. This general attitude among students in this study is mirrored in current scholarly writing on the global trend toward Internet use to serve personal and social interests. There is recent research indicating the variable uses of ICTs around the world, and how many countries are integrating ICTs and social governance. As Welman and co-authors explain in The Social Affordances of the Internet for Networked Individualism (2003):

> The move towards a networked society creates interesting possibilities for governments more used to dealing with hierarchies of local solidarities. No longer are communities local, all encompassing, and stable. Instead, people have multiple, shifting sets of localized ties. The local becomes only one kind of special interest. Even more than in the past social mobilization will be apt to develop over non-territorial issues, be it shared affect (i.e., ecology, Islam) or shared material interests.

A study by Nicole Ellison, Charles Steinfield, and Cliff Lampe (2007) titled, *The Benefits of Facebook Friends: Social Capital and College Students' Use of Online Social Network Sites* shows that undergraduate students use *Facebook* in and out of the classroom to strengthen relationships formed at school. The results of the MSU study state that, "*Facebook* members report spending between 10 and 30 minutes on average using *Facebook* each day and report having between 150 and 200 friends listed on their profile. We see that respondents also report significantly more *Facebook* use involving people with whom they share an offline connection—either an existing friend, a classmate, someone living near them, or someone they met socially," (Ellison, Steinfield, & Lampe, 2007). Students use ICTs to connect with those around them and maintain the theory of social networking by placing high importance on their relationships with others in their "network" (YorkU.ca., 2010). They ensure the linkages of these close "nodes and ties" with their families and friends by using ICTs to stay in touch while in school, and they forge new relationships with classmates, peers, and professional contacts using the same technologies (YorkU.ca., 2010).

In terms of the use of ICTs in education and learning environments, several studies show the potential and expected success of these implications. In the Australian literature review by Newhouse (2002), he finds that:

> When internet communication technology is used to support learning, it is intended that this should increase the engagement of students and in most cases increase their independence, so that students are not only required to use ICT competently but may also be required to adjust to changes in their role. That is, using ICT has implications for students beyond their ICT literacy into their perceptions of and preferences for their roles as learners. (p. 51)

As shown by this research study, undergraduate student are "plugged-in" and connected. They use ICTs with phenomenal literacy rates, in many different areas of their life, and because they are such versatile users, the implementation of ICTs in student learning and educational environments will produce progressive and positive, value-laden results (Newhouse, 2002).

RECOMMENDATIONS AND IMPLICATIONS FOR EDUCATION

The importance of personal and social aspects of Internet use are especially salient in educational research where numerous researchers have studied the potential uses for

ICTs in Education. In particular, SNSs like *Twitter* and *Facebook* have been used by universities as SNSs to connect with their students. One example of *Twitter* use within a classroom community, shows the student benefit experienced when following their professor on *Twitter*.

> Once students started twittering I think they developed a sense of each other as people beyond the classroom space, rather than just students they saw twice a week for an hour and a half. This carried with it a range of benefits, from more productive classroom conversations (people were more willing to talk, and more respectful of others), and also ... to understand what type of students they were. (Academhack. com, 2008)

This type of information shows that both professors and students can glean important educational benefits from interacting with each other in online communities.

The data gathered from this study helps provide educators with a broader picture of key aspects of student life and relationships in and out of educational settings. Primarily, it provides universities with a glimpse of just how important ICTs are to students. Students need them to gain access to their education, future job prospects, stay in touch with friends and classmates, but most importantly, to stay connected with their family. This information is important when developing appropriate technological infrastructure and policies within universities, as it pertains to student access to ICTs.

This information could also be very useful from an educational marketing perspective, as it gives universities and their stakeholders an idea of just what makes their students tick. It tells them the best way to stay in touch with them, what avenues the students will be responsive to in terms of university communication and information, and where to divert spending on marketing and developments in these areas.

Future research on ICT use and its affect on personal relationships are vast. For instance, this research study deals with data that was collected from one group of undergraduate students only, and it leaves the door open for much more research that could examine ICT influence in terms of race, gender, age, socioeconomic background and the geographical locations of friends and family, and so forth.

The significance of social technologies in daily life is evident among university students. This study and other current research illustrate how ICTs are of special importance as a means to stay in contact with friends and family. Currie (2009) states, "Specifically, during the university years, a key transition point, friendship becomes highly significant, with closer and deeper relationships being established than at previous points in young people's lives." Currie (2009) explored how *Facebook* helps strengthen students' relationship with their peers and parents. ICT popularity among university students has not dwindled. One of the reasons is that the life of a university student can be very demanding and ICTs provide students with the opportunity to maintain relationships while engaged in their studies, especially those living away from home. Not surprisingly, *Facebook* has also attracted many parents who are able to remain in contact with their children at university. As more parents become comfortable with ICTs, there is reason to believe that the influence of ICTs in maintaining personal relationships will continue to expand among the student population.

Despite the many benefits of technology to students for personal, social, and educational use, it is important for educators and designers to be aware of the dangerous risks so that interventions are possible. Students in this study also had negative experiences and views of ICT influences on their relationships which are mirrored in current research. For instance, Pierce (2009) discusses how the use of ICTs gives individuals the opportunity to replace face-to-face social interaction and how social anxiety can be increased with ICT use where individuals rely on technology too much for conversing instead of engaging in face-to-face conversations. Punamäki (2009) also note how ICT use can lead to feelings of loneliness and isolation. In this context, technology then becomes a social crutch that educators must be aware of, along with how to detect the signs of students who are at risk.

CONCLUSION

Based on the findings, there is evidence to suggest that undergraduate students do tend to use ICTs to maintain and develop personal relationships in their lives and that the use of ICTs does affect the way in which they "approach" relationships, specifically by leveraging the ease of staying in contact, and helping them form stronger and more meaningful relationships throughout the course of their university career.

Many different factors contribute to the use and dissemination of ICTs in the University of Ottawa's undergraduate students lives. Most important, however, is the fact that the majority of the participants in this study use ICTs on a regular basis, and also feel that the use of ICTs help them to maintain and develop strong personal relationships. As one female student finds:

Information and communication technologies have definitely made staying in touch with friends and family much easier, especially for me as a student living away from home. MSN, *Facebook*, and text messaging are my main means of communicating and maintaining my relationships with those that live far away and in some cases even with those that live close. (student excerpt)

The findings also indicate that the overwhelming majority of students find ICTs to be a beneficial addition to their life, despite the sometimes increase in caution while using the Internet, or the feeling of always being "plugged-in," as one female student explains:

I am an avid user of text messaging, however I am regretful of the fact that I do not speak to my friends and family on the phone or in person as much as I would like to. In one way, it is nice that I can talk to them a little bit each day. However, I feel like I am constantly hiding behind a screen. Being always 'plugged-in' has made my life a little more stressful because I easily feel like I am missing out on something by having my face constantly staring at my phone screen. (student excerpt)

It is possible that students who find ICTs to be more of a hindrance to their life than a benefit, are most likely to feel this way not because of the actual negative influence of the ICT in and of itself, but rather they feel that ICTs create less personal and intimate relationships between themselves and their loved ones.

KEYWORDS

- **Collage Findings**
- **Communication technologies**
- **Social networking sites**
- **Telecommunications**
- **Web-based learning**

APPENDICES

Table 1. Positive perceptions of ICT influence on relationships.

Theme	Theme
• Make distance obsolete between people. • Keep in touch with family and friends. • Stay connected with people. • Keep up with family and friends. • Develop more relationships with people. • Improves communication in long-distance relationships. • Connect, disclose information. • Strengthen relationships for university work. • Strengthen family relationships (*Skype, Facebook*) and friendships. • Affect private life by being always available to communicate. • Stay connected anywhere with anyone. • Connect with and family bridge gaps. • Develop more new relationships. • Can be more bold in communication, and have more time to think it over. • More playful and funny with ICTs. • ICTs are important to life. • Good for people far away. • Staying connected makes relationships healthier. • Text, Blackberry, and MSN, can move relationships along a timeline and be unobtrusive. • MSN, *Facebook*, and text to communicate and maintain relationships. • Immersed in technology—quick and easy. • ICTs are instantaneous. • Feel braver and more open.	• Use ICTs to talk to friends through chat, because it is free. • Allows people to communicate from far away. • ICTs create more dynamics for relationships and makes them easier to manage. • ICTs make communication more open and comfortable even when busy. • ICTs have always been around, cannot imagine life without SNS. • Build relationship with boyfriend. • Keep in touch around the world. • Creates more bonds with people and make new friendships easy. • Put more thought into a message, deliver when sender is not available. • Use SNS, Blackberry, increase trust levels and constant communication. • Enhances long-distance relationships. • Accommodates schedules. • Ability to multi-task and always be on-the-go. • Adds new dimension to virtual communication. • Helps global communication and long-distance relationships. • Use a variety of mediums, fast, and easy. • Positive effect on communication with those we are close to and even those who we are not close to. • Relationships can exist on different levels and can be maintained more easily.

• ICTs are easy and influence life in a positive way. • Facilitate events with family and friends. • Basis of relationships with friends, family, and peers. • Grows relationships faster. • Connect around the world. • Share pictures, social networking, and make plans. • ICTs make communication easier and faster. • Extremely simple. • Stay connected. • Make new friends. • ICTs are convenient. • Stay in contact with friends around the world. • Enhanced relationships by staying connected with family. • ICTs force people to seek new information actively. • ICTs make it easy for constant contact to keep in touch with friends. • ICTs eases and increases communication channels.	• Communicating with people who are far away and even those that live close is easier and makes them closer. • Stay connected with friends, ICT are like a third arm. • Share anything with ICTs, and made relationships more open. • Easy to stay in touch with family overseas. • Increase connectivity, always in contact. • ICTs help in maintaining relationships with family. • Maintain contact and has a positive effect on relationships. • Easier to stay in touch.

Total: 65

Table 2. Negative perceptions of ICT influence on relationships.

Theme	Theme
• Masks people. • ICTs make communication more cautious. • Text can be ambiguous and you may not know whom you are talking to. • Contact with family eliminates secrets. • They "appear" to increase interactions but they do not. • ICTs threaten traditional communication. • Makes relationships feel interrupted. • Impersonal and can create a spiritual gap. • Creates barrier because parents can use ICTs. • Can make you care more about those far rather than close. • Can be misunderstood. • Face to face for personal is more real. • Lost opportunities to strengthen relationships. • Can become dependent on the online world and disengaged.	• Prefer face-to-face, as relationships are important. • No time for oneself. • Text messages do not always truly reflect the non-verbal cues. • Prefers traditional methods of communication—eliminate the medium. • Feels constantly plugged-in, increases stress. • Hinders intimacy—broken down walls of openness. • Not as comfortable with being constantly available. • More cautious in writing. • Meaning can be lost. • Can rely too much on ICTs and create information overload. • Communicate all the time with someone, could create jealousy. • May be less personal and not as serious. • Increases communications with casual acquaintances rather than closer friends. • Relationships are text based, rather than speech. • Only trust people met face to face. • Alienate people close by. • Avoid human contact.

Total: 31

Table 3. Mixed perceptions of ICT influence on relationships.

Theme
• Can cause people to pre-judge one another, but helps professionally.
• ICTs enable life, repair broken relationships, but self-selective in relationships.
• Good social platform, but creates vague or weak plans.
• Increases relationships but not improved them.
• Furthers relationships on online meetings, but remains cautious online.
• Extends relationships, but can cause confusion.
• Nurture and grow relationships, but can cause people to not care about their personal relationships.
• Better for long-distance relationships, and helps form new ones, but harms those close by.
• Fast, simple, it is everywhere, but can be confusing and less personal.
• Available at all times, but makes communication more cautious.
• Blackberry, stay connected, but must always be reachable.
• ICTs can bring people closer, but it is not always functional.
• ICTs are constant but not a replacement for discourse in person, eliminate time/space.
Total: 13

Chapter 10

Inside Student 2.0: Student Perspectives on Navigating Online and Offline Identities

Rocci Luppicini and Patricia Barber

INTRODUCTION

This chapter draws on phenomenological research methods and collage inquiry techniques to conduct a collaborate self-study of university students views of their online and offline identities. Both visual and text-based data were included to capture the multifaceted nature and complexity of identity representation. The research question guiding this study is as follows: How do university students compare their online and offline identities? Fifty-seven student participants created personal narratives and collages comparing their online and offline identities. Findings revealed that students have a complex approach to online and offline identities that aligned with their conceptualization of online and offline identity as being either the same (continuous) or different. This division in student orientation toward their online and offline identity appeared to influence student online attitudes and behaviors in important ways. This research found key links between student identity orientation, the influence of anonymity in influencing online behavior, and online identity management strategies among students. Educational implications and future research directions are discussed.

> In this electronic age we see ourselves being translated more and more into the form of information, moving toward the technological extension of consciousness.
>
> —Marshall McLuhan

> Online I feel powerful and connected to the world 'around me' My connections allow instant communication and sharing. I don't feel restricted by rules or money so I collect the things I want and need. Some think I'm immoral but I feel my online intelligence is working to bring about a new economy. Offline I see myself as a 'unique,' bound by the restrictions of time, money, and societal values; just another cog in the capitalist machine—just like everyone else.
>
> —Student Excerpt

It appears that some of the predictions of Marshall McLuhan are becoming entrenched in the public eye as technology becomes more and more integrated into life and society. This is particularly salient among youth who are growing up within this digital age. In researching the use of mobile phones among youth, Carroll, Howard, Vetere, Peck, and Murphy (2002) indicated, "It was clear in this research that young people are adopting a lifestyle rather than a technology perspective: They want technology to add value to their lifestyles, satisfy their social and leisure needs and reinforce their group identity." This deep connection between technology and life is furthered with the growing popularity of social networks, such as *Facebook*, where young people are connecting in a significant way. In a study of *Facebook*, noted that, "reported usage levels suggest OSNs [Online Social Networks] have become firmly integrated into the communication preferences of young Canadians" (Levin et al., 2008, p. 23). Similarly, in a study conducted by the Office of the Privacy Commissioner of Canada (2009), it was found that there is the expectation among young people to be connected. "*Facebook* is the only way we communicate now. All of my friends are on now and if they aren't, they're overlooked" (Office of the Privacy Commissioner of Canada, 2009, p. 6).

A growing body of research explores how information and communication technology (ICT) and social networking sites (SNS) or OSNs are closely connected with communication, life, and identity among youth living in this electronic (digital) society. Boyd and Ellison (2008) observed that, "SNS constitute an important research context for scholars investigating processes of impression management, self-presentation, and friendship performance" (p. 219). However, there are serious gaps in the research revolving around the relationship between online and offline identity among Internet users, especially among students who, unlike previous generations, have grown up with online technology at home and within their schools. To this end, this research aims to address this gap and shed light on how students compare their online and offline identities.

This chapter draws on phenomenological research methods and collage inquiry techniques to conduct a collaborate self-study of university students views of their online and offline identities. The research question guiding this study is as follows: How do university students compare their online and offline identities? Fifty-seven student participants created personal narratives and collages comparing their online and offline identities. This chapter reports on the study findings and discusses implications for education.

BACKGROUND

In the study of identity, a number of relevant theoretical perspectives can be drawn on to help situate the study and study findings. Various identity theories and theoretical perspectives on visual research are particularly useful when examining identity which has both visual and not visual components. Leading theoretical perspectives that apply to study of online identity include, *Identity Theory, Public Commitment Theory,* and *Self-Presentational Theory.*

Identity Theory is premised on the idea that individuals have different identities for different situations. As stated by Stryker and Burke (2000), "To refer to each group-based self, the theorists chose the term identity, asserting that persons have as many identities as distinct networks of relationships in which they occupy positions and play roles." *Identity Theory* incorporates the importance of social roles, which "are expectations attached to positions occupied in networks of relationships" (Stryker & Burke, 2000, p. 286). Further, *Identity Theory* holds that within each person there is a "salience hierarchy" of identities that will be invoked given different situations and persons. Finally, the concept of "commitment" is an important element of Identity

Theory, which "is measurable by the costs of losing meaningful relations to others, should the identity be forgone" (Stryker & Burke, 2000, p. 286).

Public Commitment Theory assumes that individuals become committed to self-presentations they make public. The theory is premised on the notion that public performance represents a component of identity that increases "the individual's sense of commitment to the performed identity" (Gonzales & Hancock, 2008, pp. 172–173). Gonzales and Hancock studied public commitment to determine "whether being public online magnifies the effect of the relationship between self-presentation and identity" (2008, p. 168), that is, does an identity shift occur when online?

Self-Presentational Theory examines how individuals present themselves to others:

> Self-presentational theory is based on the assumption that in social situations, individuals attempt to control images of self or identity-relevant information (Schlenker, 1980). In the presence of others, individuals strive to portray themselves in a particular way that serves their best interests and is appropriate to the situation. The desired result is to gain approval or avoid disapproval from others (Edelmann, 1987). (Stritzke, Nguyen, & Durkin, 2004, p. 2)

This theory focuses on the self-interest of the individual to project the self in a manner that influence the participants in the social situation to have a positive perception or opinion of that person. These abovementioned identity theories (among others) are relevant to the study of online and offline identity.

Leading theoretical perspectives on visual research are also relevant to the study of offline and online identity where visual images play an important role. In contrast to written texts (such as essays), and spoken research (such as interviews), "visual research focuses on nonlinguistic images. Pictures may be used as a source of data, as a method of data analysis, and as a means of data representation (Siegesmund, 2008). There are many reasons why visual research is appealing and appropriate in the study of online and offline identity. Visual research represents what Butler-Kisber (2008) describes, "[as the] search for more embodied and alternative representational forms where meaning is understood to be a construction of what the text represents and what the reader/viewer brings to it, and the realization that we live in an increasingly visual/nonlinear world." In other words, it can be used by researchers to explore deep meaning on personal topics (like identity) while providing additional data (visual) to better aligns with the multimedia reality of online and offline identity experienced by individuals.

METHOD

This chapter draws on phenomenological research methods to describe meaning for individuals of their lived experiences of a concept or a phenomenon (Creswell, 2007, pp. 57–58). A phenomenological study was selected for the research on identity because the goal is to identity the common experiences of university students using computer-mediated communication.

The chapter also draws on collage inquiry techniques to provide a rich multi-faceted examination of students views on their online and offline identity following

Butler-Kisber (2008). Butler-Kisber suggests a "methodology when analyzing collage, where "found" materials that are either natural or made are cut up and pasted on some sort of flat surface" (2008, p. 266). Butler-Kisber's process is as follows:

1. Create a series of collages in response to a research question.
2. Give collages a title to identify the essence of the image.
3. Discuss with other researchers, preferably working on the same area of inquiry.
4. Identify the "intuitive facets" of each collage.
5. Look "at the colors, shape, content, and composition (Rose, 2001), and [find] commonalities across the collages, [and] new conceptualizations" (2010, p. 116).

The emphasis on finding commonalities is of particularly importance as the theoretical perspective applied is phenomenological and a key outcome is to determine central themes "which expresses the essence of these clusters" (Groenewald, 2004, p. 20). This logic in research design is intended to encompass both text and collage data to provide a richer portrait of how students navigate their online and offline identities.

In terms of data collection, 57 student participants from an undergraduate research course in communications created collages and narrative responses to the following question: How do my online and offline identities compare? Next, students e-mailed their assignments to the principle researcher for posting to a public research blog. The principle researcher (who was also the course instructor) obtained the written consent from those students who agreed that their texts and collages could be used for other research purposes.

Data analysis involved horizontalization to every significant statement relevant to the topic to help organize and creating meaning from the data (Creswell, 2007). More specifically, the researchers conducted a line-by-line examination of the texts to identify if the respondent expressed having different or similar online and offline identities and grouped participants' texts into these categories while recording any observations that cut across the different groups. Next, the researchers conducted an examination of the collages using a framework that observed the visual data using the following elements: divisions, face, body, how the body was portrayed, words, and other images. Similar data analysis steps were applied to both text and collage, "preparing and organizing the data (i.e., text data as in transcripts, or image data in photographs) for analysis, then reducing the data into themes through a process of coding and condensing the codes, and finally representing the data in figures, tables, or a discussion" (Creswell, 2007, p. 148). Finally, the researchers examined both text and collage data to develop a composite summary of online and offline identities.

FINDINGS

Text Findings

The key findings from the text analysis is summarized and divided into six main thematic areas, namely: (1) similar versus different identities, (2) managing online identity (3) information privacy and anonymity, (4) claiming new identities, and (5) challenges in managing identities.

Similar versus Different Identities: Findings indicated that the majority of participants viewed their online and offline identities as different. The research found that the majority of participants 58% (33/57) self-identified as having a unique online identity that was not the same as their offline identity. Participants were grouped based on whether they were more open or more reserved online, or whether they did not seem to fit into any of these categories. Some participants were more open online. As stated by one participant, "I feel that my online identity can sometimes be a bit more outgoing than my offline identity. I also find that I attempt to portray my online identity in a more colorful way" (student excerpt). A second student noted, "My offline and online personalities are somewhat different from each other, as I try to present myself as a more outgoing, sociable and carefree person in my online profiles " (student excerpt). In contrast, some students revealed that they were more reserved online. As stated by one student, "When I'm online I tend to be a bit more reserved and keep a lot of information hidden because of safety issues" (student excerpt). Another student expressed a similar sentiment, "Offline I am extremely extroverted and expressive. Online however I am weary of what I say because you can never ensure who is on the other side of the screen you are talking to" (student excerpt).

For other students there was less certainty concerning their online identity which they minimized in importance compared to their offline identity. Two student extracts illustrate this point:

Online, I am content with visiting my favorite websites and resuming living by my REAL, offline identity. In comparison, I do not importance on my online identity, whereas I believe my offline identity is my only identity. I do not live through my *Facebook* page. (student excerpt)

I feel like my offline identity differs a lot from my online ability, in that my offline is definitely real, whereas the online me, is a fake and fabricated version of me. (student excerpt)

Findings also revealed that 42% (24/57) of the participants self-identified as having similar online and offline identities. Extracts from responses reflect an awareness of certain nuances between the two:

My online identity is just a virtual extension of my personality, however it is an 'edited' version of me. (student excerpt)

I am myself online and offline, what you see is what you get. (student excerpt)

My online identity is my offline identity. I prefer that my online and offline identities are one in the same. (student excerpt)

Because I am not engaging in any anonymity, I do not feel my identities change drastically. (student excerpt)

A number of the participants rationalized the similarity between their offline and online identity by stating that their online network was comprised of family or people they know. As stated by one participant, "I think that my online and offline personalities are similar. Since I am usually only interacting online with people that I know [offline]" (student excerpt). Another student added, "Since I only ever talk to people I know online, my online and offline identities are very similar" (student excerpt).

Based on the findings, it appears that students tended to avoid online anonymity (at least in part) to preserve the continuity of their online and offline identity.

Managing Online Identity: Managing their online identity emerged as a second theme among participants. Findings indicated that the management of online identity was a major theme for participants. 32% (18/57) of the participants communicated in their texts an awareness of managing their online identity in a variety of ways. Participants managed their online identity in a number of ways. First, some participants managed their online identity by presenting a positive self image. One participant indicated, "I want to share my personality and character online in a way that defines me in a positive way. It is important to me to leave a digital fingerprint that I am proud of, and that truly represents me" (student excerpt). Other responses reflected an effort for students to go beyond their offline selves in their online presentations of self:

I attempt to portray my online identity in a more colorful way. (student excerpt)

I allow myself to appear more energetic than I come off in person. (student excerpt)

I try to present myself as a more outgoing, sociable and carefree person in my online profiles. (student excerpt)

Second, some participants manage their online identity by carefully managing how they write about themselves and what they write to others. Second, some participants described controlling their image through writing as a self-management technique. As stated by one participant, "When I communicate online the things I say are filtered because I take a second to think before I press send" (student excerpt). A second student noted, "I am careful with what I say to others online, as emotions can be misinterpreted through typing" (student excerpt). In a similar vein, a third student expressed, "On my *Facebook*/online identity, I probably appear very confident and that is because of the rhetoric I've used to describe myself, and in person I am also confident as well, but it is easier to see that confidence through my online identity" (student excerpt). Other participants engaged in other types of management strategies to optimize their online and offline identities. Third, a number of other self-management strategies were offered by students. One student noted, "When I'm online; I'm aggressive, I enhance the appearance of my good qualities while toning down my poor qualities" (student excerpt). Another student stated, "I have the chance to carefully think about what I share with the online world, and manipulate it to my advantage. It's almost as if I am advertising myself for the world, so you censor yourself to show your best qualities" (student excerpt). A third student conveyed a similar sentiment, "Since information on the web is eternal, I chose carefully which image to portray" (student excerpt).

Information Privacy and Anonymity: Managing personal information and privacy emerged as a third theme. It was found that 19% (11/57) of participants communicated that they protect their personal information or privacy when online. As stated by one student, "I always make sure to not convey a lot of personal detail in my online identity, for safety reasons" (student excerpt). Another student added, "Online the information I am likely to provide others is more superficial" (student excerpt). A number of other statements expressed similar views concerning privacy:

I don't share my private information or post any personal information on social networking websites. (student excerpt)

I am concerned with protecting my privacy, and exercise caution when defining my online identity. (student excerpt)

I am very private online. I don't divulge information unnecessarily. (student excerpt)

Anonymity issues were discussed by a number of students regardless of whether they perceived themselves as being more or less open online compared to their offline conduct. The following are three extracts from the texts that are influenced by the anonymity of online communication to be more open:

My online identity is fearless because I can hide behind a computer screen, however offline I have nothing to hide behind and keep myself at more of a distance. (student excerpt)

I feel confident in who I am, I will engage with others because I can hide behind the screen. (student excerpt)

I can be more comfortable with myself and how I behave, because there is a shield of anonymity. (student excerpt)

Students that indicated that they were more reserved online were also concerned with the anonymity of online communication:

I cannot bring myself to share an opinion on-line. I am more afraid of what strangers might say to me then a group of my close friends. (student excerpt)

Online however I am weary of what I say because you can never ensure who is on the other side of the screen you are talking to. (student excerpt)

It is interesting to note how anonymity is used by students to explain both their reluctance and motivation to express themselves online. It appears that student perceptions of anonymity are strongly linked to how students navigate their online and offline identities. That is, students that tend to view their online and offline identity as continuous tend to be more reserved online and avoid personal information disclosure within anonymous online interactions. Conversely, students that tend to view their online and offline identity as different, tend to be more open online and not avoid personal information disclosure within anonymous online interactions. This will be discussed further in the chapter discussion.

Challenges in Managing Identities: Another theme identified in this study by numerous participants revolved around identity management challenges which appeared in 15% (9/57) of the texts. Balance and consistency were concerns for a few participants. As stated by one participant, "Overall, I try to keep my offline and online identities balanced, and I always try to portray most aspects of my personality evenly in order to stay true to myself" (student excerpt). A second student stated, "I make a conscious effect to try and maintain a consistent identity in both spectrums" (student excerpt). The challenge of balance and consistency mainly affected those holding the belief that their online and offline identities should be consistent. One student described the experience as being caught within two worlds, "Often caught up in between the two distinctive worlds, a sense of stress sets in. Trying to manage, double identity, while making sure they coincide proves to be difficult" (student excerpt).Often that challenge can be overwhelming as noted by another student, "I find I keep my guard up most of the time, but sometimes I simply feel overwhelmed online" (student excerpt). The challenge of managing online and offline identity is one that is often a surprise to the individual, catching him or her off guard. As on student recollected, "Sometimes it is shocking to me how my personality is at polar opposites depending on whether I am online or in a human 'offline' situation" (student excerpt). Another participant added, "In somewhat of a surprise, I have been told that my two personalities can essentially be called one, as the differences are minute" (student excerpt).

Claiming New Identities: The study also uncovered how some participants have discovered new parts of themselves that do not come out in their offline world. A few respondents, 7% (4/57) indicated that their online identity influences their offline identity, in most cases positively. One student stated, "I find that the online world inspires me to be more outgoing and unreserved in the offline world" (student excerpt). This theme was shared by others as reflected in this statement, "I feel that my online identity is the person I want to be, while my offline identity lives the life I love but feel could be much more if I could implement the skills I utilize online" (student excerpt. The attraction of online identity appears to surface in these statements that acknowledledge the value-added aspect of online identity to the lives of students.

Collage Findings

The following elements emerged when examining the collages: divisions, words, and images.

Divisions: This research found that a majority of the collages used divisions when answering the research question: 71% (17/24) of the collages expressing similarities were divided; and 70% (23/33) of the collages expressing different online and offline identities were divided. It is noteworthy that the divisions between collages expressing online and offline difference often appeared to be more contrasted. Common divisions were created using horizontal, vertical and diagonal lines:

| Similarity | Similarity |

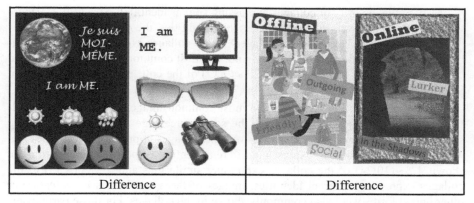

| Difference | Difference |

Words: The majority of the collages used words and images. The most frequent words on the collages were "online" and "offline." On many of the collages these particular words were used as titles. Also, in a number of collages, words were selected and patterned in such a manner that ambiguity in the message was significantly reduced. This is illustrated in the collage example below where the words "I shouldn't change" appear to support the value of the student to maintain similar online and offline identities:

| Collage 5 |

Also, in the majority of collages expressing similar identities, 61% (20/33), there were no titles. However, in those collages that used titles, the most common title was "online" and "offline" (12% of the collages). In the collages expressing different identities, 48% (16/33), there were no titles. And, the most common title was "online" and "offline," which was incorporated in 53% (9/17) of the collages expressing different identities.

Images. Faces and bodies were the most common images identified in the collages. Faces were shown in relatively the same percentage in collages expressing similarities and differences: 46% (11/24) of the collages expressing similarities and 45% (15/33) of the collages expressing different identities. Bodies or body parts were shown 42% (10/24) of the collages expressing similarities and 57% (19/33) of the collages expressing different identities. Images of faces and bodies in caricature or other unreal forms were also commonly shown in the collages. For those participants who expressed similar identities, such images were shown on 25% (6/24); and for those participants who expressed different identities, such images were shown on 32% (10/31). It is particularly interesting to note that in some collages images were selected for comparative purposes. For example, the different cats shown on each side of the collage on the right expressed the "similar" nature of the identity (the images are in the "cat" family). And, the contrast between the white and the black cats in the collage on the left is used to express the essential differences in identity.

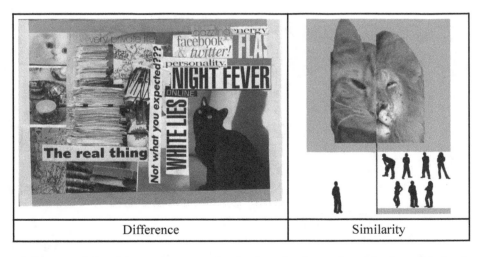

| Difference | Similarity |

In general, participants were creative in the selection and positioning of their visual messages using a variety of images in different ways. The following is an example of such creatively with the accompanying text that explains the meaning of the images used.

 Overall, my online and offline identities are fairly similar (symmetrical background image). There are, however, a few differences. My online identity is represented by a more cautious self. I am careful with what I say to others online, as emotions can be misinterpreted through typing (eggshells). Online, I am anonymous, making me more forward and inquisitive about other people's lives (Curious George). My offline identity is fairly open. My outgoing personality makes it easy for people to approach me and learn about me (open book). Like my online identity, I am cautious to what I say to others (whispering).

INTERPRETATION

The key findings discussed in this section are divided into six areas, namely:

1. Complexity of Rationale for Choosing Online/Offline Identities,
2. Identity that is performed in Different Environments,
3. Overcoming "Gating" Challenges,
4. Performing or Not Performing in an Anonymous Environment,
5. Protecting Personal Information/Privacy,
6. Managing Impressions Online, and
7. Claiming new Identities.

Complexity of Rationale for Choosing Online/Offline Identities

This research found that individuals have either similar or different online identities and their choices are based on different reasons which reflect a degree of complexity. Merchant notes this complexity: "Whilst it is relatively easy to accept that we draw on different facets of our multiple identities in different communicative settings (both on- and offline), the actual choices we make are more complex and certainly more interesting" (Merchant, 2006, p. 239). This complexity is illustrated when comparing the decision of one respondent to be more open online with another respondent who chooses to be less open. The participant's decision to be more open online is based on knowing they cannot be rejected, "Online I feel more willing to share, knowing I can't be rejected" (student excerpt). Conversely, another participant's decision to be less open online is based on being "more afraid of what strangers might say to me [online] than a group of my close friends" (student excerpt).

Merchant argues that "[a]lthough choices are available, they are constrained by the affordances of a particular medium and the social conventions that emerge around that particular form of communication. So outside of the game environment we would anticipate that most of the time, online communication would involve some enactment of real-world identity; or perhaps, more accurately, real-world identities"(Merchant,

2006, p. 237). This research found the enactment of real-world identity particularly salient for participants that had similar online and offline identities, as shown in the following responses:

> I think that my online and offline personalities are similar. Since I am usually only interacting online with people that I know (student excerpt)

> I try to keep my internet personality as close as possible to my real personality... I usually only communicate with people that I know personally so I have nothing to hide. (student excerpt)

> Since I only ever talk to people I know online, my online and offline identities are very similar. (student excerpt)

Online relationships that are based on offline relationships are known as "anchored relationships" (Zhao, Grasmuck, & Martin, 2008, p. 1818). Ellison, Steinfield, and Lampe's work highlight this type of relationship in their study of the use of *Facebook* to maintain offline connections and "social capital": "The strong linkage between *Facebook* use and high school connections suggests how SNSs help maintain relations as people move from one offline community to another" (Ellison, Steinfield, & Lampe, 2007, p. 1164). Maintaining connections with offline connections using *Facebook* was also found in qualitative research conducted by Decima Research:

> Consistently, participants named keeping in touch with friends as their most positive experience on, or benefit from, Facebook. Many told stories of reconnecting with long lost friends or people met while travelling and being easily able to stay in touch on Facebook despite vast distances and time zones separating them. (Office of the Privacy Commissioner of Canada, 2009, p. 10)

Merchant also reflects on the link between online and offline networks: "For much computer-based communication the imagined audience relates quite closely to actual social networks. Of course this overlap is more the case in some environments (such as email and instant messaging) than others (such as web pages and MMPOLGs [massively multiplayer online games])" (Merchant, 2006, p. 240).

The particular influence of linked online and offline networks—anchored relationships - on identity is reflected in the *Theory of Identity*, which was not intended for online networks, but, rather for offline "real" social structures; however, the concept of commitment in Identity Theory seems to apply to those users that operate within closely linked online and offline environments. "Commitment" is explained as follows:

> Persons ... tend to live their lives in relatively small, specialized networks of social relationships. Commitment refers to the degree to which persons' relationships to others in their networks depend on possessing a particular identity and role; commitment is measurable by the costs of losing meaningful relations to others, should the identity be for gone. (Stryker & Burke, 2000, p. 286)

Stryker and Burke's research focus on the influence of social structures on identities and "self-verification" and how "self-verification" develops and maintains social structures. Their research is important since it highlights the important influence of small and linked online and offline networks, which are "social structures," in facilitating similar online/offline identity. This lends support to Identity Theory as a suitable explanation for this aspect of the chapter findings.

Identity that is Performed in Different Environments

Another interesting finding from this study links to the performance dimension of identity as a social product. Resent research explains how identity can be a social product:

> Our results suggest that identity is not an individual characteristic; it is not an expression of something innate in a person, it is rather a social product, the outcome of a given social environment and hence performed differently in varying contexts. Depending on the characteristics of the environment in which they find themselves, individuals will choose to claim identities that can help them better situate within the given social environment. 'True selves,' 'real selves,' and 'hoped-for possible selves' are products of different situations rather than characteristics of different individuals. (Zhao et al., 2008, p. 1831)

This perspective is different from that of Identity Theory since Zhao et al. entertain the possibility of being in different social environments and responding or managing the environment by "claiming identities" appropriate to the situation. This seems to be the situation mainly for participants with different identities, however, it may also apply to some participants who expressed similar identities, given the qualifications they expressed for their "similar" identities. The following extracts, one from each category of identity—different and similar—highlight the influence of identity performance in the online environment, which appears to be viewed as representing a less judgmental social situation:

> While online I have the power to be whoever I want to be and do not have to worry about being judged. (student excerpt)
>
> I have many shared interests between both identities although some things are only for one of my identities. My online identity is more expressive and opinionated, yet offline I have somewhat more reserved. I feel sometimes, I can express myself more freely online because there is no one judging me or singling me out. (student excerpt)

It appears, as well, that for many participants, there is an acceptance that their everyday reality is that they must perform identity in a manner suited to the environment. This is reflected by Zhao et al.: "In the Internet era, the social world includes both the online and offline environments, and an important skill people need to learn is how to coordinate their behaviors in these two realms" (Zhao et al., 2008, p. 1832).

This research found that a few participants are shocked and disappointed by the differences between their online and offline identities, and this represents the reflexivity on the part of the participants to their reality. The following are extracts from these comments:

> Sometimes it is shocking to me how my personality is at polar opposites depending on whether I am online or in a human 'offline' situation. (student excerpt)
>
> Unfortunately, my online identity does not match well with my offline identity. (student excerpt)

The findings in this research also found that for a few participants the reality of coping with performing with more than one identity is challenging. The following extracts reveal this challenge:

> I find I keep my guard up most of the time, but sometimes I simply feel over-whelmed online. (student excerpt)
>
> Often caught up in between the two distinctive worlds, a sense of stress sets in. Trying to manage, double identity, while making sure they coincide proves to be difficult. (student excerpt)

Such difficultly is also reflected in existing research which suggests that there is increased difficulty in dealing with a greater number of identities: "the greater the number of related identities, the greater the difficulty of dealing simultaneously with relationships among them. There is no clear way of attacking the issue at hand (Stryker & Burke, 2000, p. 292).

Overcoming "Gating" Challenges

Shyness and physical appearance are only two of many "gating" challenges that are evident in face-to-face interactions, but not evident in online interactions where the interaction is text-based. The research notes the "identity empowerment" that happens online for individuals that would be considered to have "gating" challenges:

> Research has shown that the removal of physical 'gating features' (stigmatized appearance, stuttering, shyness, etc.) enables certain disadvantaged people to bypass the usual obstacles that prevent them from constructing desired identities in face-to-face settings. The emergent online anonymous environment also provides an outlet for the expression of one's 'hidden selves' (Suler, 2002) and the exploration of various non-conventional identities. As such, the Internet plays an important role in identity empowerment. (Zhao et al., 2008, p. 1818)

This research found that "identity empowerment" was a key element for participants who self-identified as shy or commented on their physical self. The following extracts highlight this online empowerment:

> Offline I am more shy and reserved. ¼ When offline I am an observer first and then I jump into a situation. When online however, I am more likely to act first and think later. (student excerpt)
>
> Online I can be a thin attractive female. ¼ I am a shy person offline and enjoy wearing sweatpants, not caring what I look like. (student excerpt)

Other research as well supports online confidence for shy individuals: "In sum, this study represents one of the first systematic investigations of the relationship between shyness and CMC use. The findings indicate that the absence of visual and auditory social cues online facilitates less inhibited social interactions for shy individuals" (Stritzke et al., 2004, p. 17).

Performing or not Performing in an Anonymous Environment

Anonymity surfaced in this research from two perspectives: Some participants viewed it as an opportunity to be open; and other participants viewed it with fear. With respect to the former—not fearing anonymity—the following extracts highlight the benefit of hiding behind the screen when presenting a confident identity:

> My online identity is fearless because I can hide behind a computer screen. (student excerpt)
>
>   I feel confident in who I am, I will engage with others because I can hide behind the screen. (student excerpt)
>
>   I can be more comfortable with myself and how I behave, because there is a shield of anonymity. (student excerpt)

Such responses are consistent with Suler's perspective on the positive aspect of anonymity as a way of managing multiple selves:

> The desire to remain anonymous reflects the need to eliminate those critical features of your identity that you do not want to display in that particular environment or group. The desire to lurk—to hide completely—indicates the person's need to split off his entire personal identity from his observing of those around him: he wants to look, but not be seen. Compartmentalizing or dissociating one's various online identities like this can be an efficient, focused way to manage the multiplicities of selfhood. (Suler, 2002, p. 459)

Joinson's research also suggests positive aspects of anonymity for increased self-disclosure. Self-disclosure is defined as the 'act of revealing personal information to others' (Joinson, 2001, p. 178). Joinson's work found that "self-disclosure is higher in CMC than FtF, and that both visual anonymity and heightened private/reduced public self-awareness can be implicated in this effect" (Joinson, 2001, p. 190).

However, this chapter found that self-disclosure was not the norm for some participants. Rather, personal information and identity attributed presented by individuals were censored. The following texts highlight this:

> Online the information I am likely to provide others is more superficial, it is "above ground" for the most part. (student excerpt)
>
>   Online I am caged—it is not my personal traits you get to know, but rather what I do, what I watch, and what I enjoy. It is a portrait of me, with lots of exclusions. (student excerpt)

Therefore, while Joinson's study indicates increased self-disclosure as a result of anonymity, this research found that, in some cases, the nature of the self-disclosure is less than full disclosure. This was particularly salient along those who valued the continuity of their online and offline identity.

Managing Impressions Online

This research found that managing impressions online was very important. In the responses provided key phrases were used to express the intentionality to manage impressions such as "I attempt to portray my online identity" (student excerpt), "I allow myself to appear" (student excerpt), and "I chose carefully which image to portray" (student excerpt). For some participants, the permanence of the information online is the reason for managing their images:

> Since information on the web is eternal, I chose carefully which image to portray. (student excerpt)

> I understand that any information or online activity can be archived, even if it is deleted so I have adopted a future oriented perspective in all of my online activity. I want to share my personality and character online in a way that defines me in a positive way. It is important to me to leave a digital fingerprint that I am proud of, and that truly represents me. (student excerpt)

The permanence, or "persistence," of information online is one of four aspects of self-presentation on social network sites that Boyd identifies that differ from the processes of offline identity formation. The three other aspects are: searchability, replicability and invisible audiences. Boyd writes about these elements:

> [I]t is virtually impossible to ascertain all those who might run across our expressions in networked publics. This is further complicated by the other three properties, since our expression may be heard at a different time and place from when and where we originally spoke. (Boyd, 2008, p. 126)

Image management is also linked to a number of factors including uncertainty—not knowing your audience and not knowing how an individual's words and images will be used or understood by others. An important implication of this uncertainty is that it can motivate people to self-present their best or most positive selves. This is demonstrated in the following texts:

> I attempt to portray my online identity in a more colorful way. (student excerpt)

> When I'm online; I'm aggressive, I enhance the appearance of my good qualities while toning down my poor qualities. (student excerpt)

> My online identity often reflects only the good and positive side of me to others. (student excerpt)

> I try to present myself as a more outgoing, sociable and carefree person in my online profiles. (student excerpt)

These findings mirror the research on image management, which indicates that, in general, online self-portrayals are positive:

> Others have found that online self-presentations tend to portray strategies of ingratiation and competence, suggesting that individuals want to facilitate social relationships while trying to impress others. In all, self presentation appears to be a highly deliberate process. In fact, Stern (2004) parallels the process of building personal Web pages to the real-world act of identity construction in that each component of the site appears to be carefully decided upon to create a particular impression of the self. (Gonzales & Hancock, 2008, p. 170)

A particularly effective means of putting a positive perspective on one's identity is the written word. This research found that for many participants there was careful consideration of the thoughts and messages and words expressed before making them public. This is reflected in the following extracts of texts:

> When I communicate online the things I say are filtered because I take a second to think before I press send. (student excerpt)

> On my *Facebook/* online identity, I probably appear very confident and that is be-
> cause of the rhetoric I've used to describe myself, and in person I am also confident
> as well, but it is easier to see that confidence through my online identity. (student
> excerpt)

The literature on writing online and creating identity for audiences supports these
findings:

> In writing online, how we conceive of and respond to audience, and how we imag-
> ine our readership, is clearly of central importance. This point is emphasised by
> Hine when she argues that the production of a web page 'is made meaningful pri-
> marily through the imagining of an audience and the seeking of recognition from
> that audience.' (Merchant, 2006, p. 240)

The findings on identity management appear to be congruent with *Self-Presentational Theory* and the idea of presenting only your best self to others for your own
benefit.

Protecting Personal Information/Privacy

For a few participants, there was a strong concern about protecting personal informa-
tion/privacy. In fact, 19% of the participants explained that their online interactions
were managed in a way to protect their personal information/privacy. The following
texts from participants reveal their interest in privacy:

> I don't share my private information or post any personal information on social
> networking websites. (student excerpt)
>
> I feel that I have more control over the information that I disclose online, which
> allows me to choose what I prefer to disclose and what I would rather keep private.
> I am concerned with protecting my privacy, and exercise caution when defining my
> online identity. (student excerpt)
>
> I try to keep a lot of information about myself off the internet. (student excerpt)

The concern for privacy was a moderately important theme in the findings where
19% of participants expressed concern for their online privacy. A possible explanation
for this is found in a Canadian government qualitative study on privacy issues and
risks to SNS, "[a] trend that emerged across age groups was a general acquiescence
that the information you put online is at risk of being used in a harmful way" (Office
of the Privacy Commissioner of Canada, March 20, 2009, p. 12).

Claiming New Identities

The findings reveal that for some participants after demonstrating certain characteristics online—gaining practice in an audience—they are motivated to take on these
same characteristics in their offline identity. The following are two comments from
participants on this transference:

> However, when I'm sitting in front of a keyboard, it's much easier for me to type
> out what I want to say. I can backspace, copy, paste, edit and save my thoughts for

another day. However my offline identity has slowly been adopting this skill and I've become more confident with speaking out. (student excerpt)

I find that the online world inspires me to be more outgoing and unreserved in the offline world. (student excerpt)

These findings are validated by the research findings of Gonzales and Hancock:

Our results indicate that making one's self identifiable on a publicly accessibly blog can generate a new self-concept based on that self presentation. .... Our findings suggest that when people walk away from the keyboard they may take with them aspects of the online self-presentation. (2008, p. 179)

These findings, therefore, are also consistent with *Public Commitment Theory* in that there is increased commitment to an identity once it has been performed publicly. And, relative to Gonzales and Hancock's research, once there has been public commitment to an identity online, identity shift can occur with the performed online identity becoming the offline identity.

Using Collages to Collect Visual Data

Visual inquiry such as arts-informed research is increasingly viewed as a source of added-value to research. Butler-Kisber explains:

There is a burgeoning interest in using arts-informed research to counteract the hegemony and linearity in written texts, increase voice and reflexivity in the research process, and expand the possibilities of multiple and diverse realities and understandings. The search for more embodied and alternative representational forms where meaning is understood to be a construction of what the text represents and what the reader/viewer brings to it, and the realization that we live in an increasingly visual/nonlinear world, have naturally led researchers to explore the potential of visual texts, collage being one possibility. (Butler-Kisber, 2008, p. 268)

Bagnoli also reflects that "[t]he inclusion of non-linguistic dimensions in research, which rely on other expressive possibilities, may allow us to access and represent different levels of experience" (Bagnoli, 2009, p. 547). Weber as well discusses the value of the non-linguistic dimensions of images is discussed in many ways:

An image can be a multilayered theoretical statement, simultaneously positing even contradictory propositions for us to consider, pointing to the fuzziness of logic and the complex or even paradoxical nature of particular human experiences. It is this ability of images to convey multiple messages, to pose questions, and to point to both abstract and concrete thoughts in so economical a fashion that makes image-based media highly appropriate for the communication of academic knowledge. (Weber, 2008, p. 43)

In this research collages were incorporated as a source of data, in addition to text. Common techniques used by students in the collages were particularly relevant to the question to which participants were asked to respond. For example, the use of lines to create divisions on the paper was frequently included to set-up the comparison between the online and offline identities. As reported in the Findings section in this paper, 17/24 or 71% of the collages expressing similarities and 23/33 or 70% of the collages expressing different identities were divided using divisions such as lines (horizontal, diagonal, vertical, colored). Other comparative techniques included using titles with the most frequent title being Offline Online, and, for greater certainty one collage included the title "Offline vs. Online." In some cases, each word in the title was associated with a particular section on the collage, which helped with the interpretation of the collage. Other collages omitted the reference to online and offline, and used titles that referred to the essence of the identity in each of the environments. For example, the title "Color talk vs. Black Out" was positioned with images that made it clear that black out was a reference to reserved and/or silent online identity. Similarly, another collage used the title "Open Restricted" and it was positioned in a manner that it was evident that the online identity is restricted since the word restricted was aligned with an image of a female at a computer. Finally, "comparative images" such as white and black cats (below left), thin and thick bodies (below right) were also used to express meaning about identities.

| Collage Sample | Collage Sample |

Collage techniques employed by students demonstrated a means of expressing personal meaning and representing identity within this study. These techniques are referenced in existing literature. For example, in her evaluation process Butler-Kisber includes colors, shape, content and composition (2010, p. 116). Composition was particularly used in the collages especially creating the divisions. Butler-Kisber provides the researcher with other areas of evaluation that were seen in this research, such as metaphors:

> A single or linear thought 'gives way to relations of juxtaposition and difference' and the fragments both work together or in opposition to produce new connections and illuminating ideas. Second, the use of metaphor (similarity or comparison) and metonymy (contiguity or connectedness) and the gaps and the spaces within a collage reveal both the intended and the unintended. (2010, p. 104)

A key interest in collages examined whether words were used, and if so, how they were used. As has been discussed earlier in the chapter, only a few collages did not use words. In one collage, the words were integrated into the images of the collage to minimize any ambiguity of the message: the message I shouldn't change support the messaging of similar online and offline identities.

Collage Sample

It is interesting to note how the use of language continued to be an important communication tool even if the collage was essential a visual methodology. This aligns with Gauntlett's (2004) article, "Why bother with words at all?," where he concludes: "In terms of academic research, or more specifically research about the ways in which individuals relate to media material, it would be difficult to ditch words altogether" (2004, p. 8).

Based on the close identification of faces with identity, it is not surprising that this study found that faces, real or caricatures, were frequently used on the collages. For example, in this research pictures of faces were used on 46% of the collages (11/24 for those collages expressing similarity and 15/33 for those collages expressing differences); and, 29% of all of the collages contained a caricature of a face (6/24 for those collages expressing similarity and 10/31 for those collages expressing differences). The use of faces is reflected in the importance of sharing photos on social network sites, which was found in the Canadian research on: "Pictures are clearly a valuable, if not vital, element for most of the users we studied" (Office of the Privacy Commissioner of Canada, 2009, p. 3).

Composite Summary of Online and Offline Identity

As demonstrated in this chapter and in other research, online and offline identities develop based on a complex interaction of factors. The question of identity is somewhat simplified for a few with similar online and offline relationships: they will maintain the same identity in both environments because of expectations of roles by family and friends. For others, the decision to have different identities appears to be based on their perspective on anonymity: For some anonymity is perceived as something to fear (the fear of the unknown audience) and they are more reserved online; and, others respond with confidence to the new found freedom offered by anonymity. For some shy individuals anonymity provides an opportunity to experiment with identities that are different from their offline identity.

It also appears that an individual's confidence with the online identity can influence an identity shift to greater congruence between online and offline identities. That is, asserting one's offline identity publicly in the online environment can facilitate the identity shift in the offline environment. Ribeiro (2009) argued that online identity be powerful when online experiences are "so vivid that the characteristics of virtual identities would gradually be embedded on one's offline life" (Ribeiro, 2009, p. 293). In particular, for students that are shy offline, online interactions expand opportunities for personal growth through online experiences because they are no longer "limited to the circumstances and social situations located in the geographical nearby spaces [and] would have their range of experiences expanded through the knowledge and contact with other realities, values, and beliefs" (Ribeiro, 2009, p. 294). In this way, online life can facilitate the development of offline life, skills and learning opportunities tested out by individuals online.

This was also reflected in a moderate number of personal narratives and collage work from students in this study.

In addition, some individuals were concerned with their privacy and tended to be cautious within online communication, particularly those that placed a high value on the congruence of their online and offline identity. For certain individuals, it was important to have consistent and balanced identities in the two environments. Moreover, some individuals found it stressful to manage identities in the two realms. These individuals appeared to be keenly aware of the audience in the online environment along with the permanence of online information presented, which motivated them to invest great effort and care in managing their identities in a manner that reflects positively on them.

CONCLUSION

This chapter drew on phenomenological research methods and collage inquiry techniques to conduct a collaborate self-study of university students views of their online and offline identities. Findings revealed student perceptions and experiences in navigating their online and offline identities. The significance of this study is twofold: First, findings draw attention to the challenges experienced by some users in managing two identities—online and offline. This provides insight to help online learning instructors and designers better connect with the students by designing online learning environments in accordance with student values. Second, the study explores the use of collage as an important data source in qualitative research, particularly with respect to the issue of authenticity and the nature of online environments which are both text and image rich. Overall, this chapter highlights how some users experience a high level of emotional discomfort in managing their two identities which instructors and online learning designers need to be aware of. This contributes valuable insight to help leverage understanding about how students navigate their online and offline identity.

DISCUSSION

Limitations

The main study limitation revolves around the issue of subjectivity when attempting to analyze and evaluation visual data. One challenge in the present analysis concerned the interpretation some collage images and trying to understand the whole message. This is consistent with the research which notes that: "[n]ot all images are equal or equally effective or valid. Images, like words, can be used to twist and distort and mislead" (Weber, 2008, pp. 49–51). For example, in this research it was difficult to determine if one particular collage was repeating the question that was being asked— How do my online and offline identities compare?—or responding to the question.

Collage Sample

To try to clarify the message, the researcher referred to the respondent's text: "My online identity is just a virtual extension of my personality; however it is an 'edited' version of me" (student excerpt). Is this what is represented in the collage? This particular issue relates to the challenge of subjectivity in the evaluations of collages. And, the research is quite clear that this is a concern in visual inquiry:

A researcher can strive to be faithful to participant stories and experiences, presenting a faithful rendering of their original form and being faithful to participant meanings and understandings. Also implicit in the goal of faithfulness is the idea of the limitations of striving—which is to say that though the researcher will strive to be faithful, the limitations presented by their individual subjectivity, however closely audited, must still be taken into account. (Galman, 2009, p. 214)

Other research suggests that co-ordinated guidance should be available to support evaluation of visual inquiry. Butler-Kisber suggests, "What is needed, however, is an integration of the criteria for evaluating arts-informed research with those for evaluating visual images, with a particular focus on collage" (Butler-Kisber, 2008, p. 273). Further, she comments that "The conversation around quality and evaluation of collage inquiry needs to continue. As more examples are shared by researchers and artists cross boundaries, the more fruitful the conversation will be" (Butler-Kisber, 2010, p. 119). Galman also suggests that "exploring and developing analytic techniques for use with a variety of image-based and other data may contribute to more widespread application" (Galman, 2009, pp. 212–213). Thus, many of the limitations discussed are getting worked out with each new study that helps to advance the clarity of the evaluation process and criteria.

Recommendations for Educational Research

This chapter applied sophisticated collage inquiry techniques to explore how university students compare their online and offline identities. This chapter provided useful insider information concerning the values and behaviors of students attempting to navigate online and online environments. More educational research should focus on obtaining the "insider" perspective on online and offline learning from the students affected by it in order to better identify their needs, interests, and barriers. It was also found that, visual inquiry techniques (like the one employed in this chapter) are especially well suited for such educational research interested in gaining personal insider information about users values and experiences of online and offline environments. To this end, the following suggestions are offered by the authors for improving on the application of visual inquiry methods:

- Instruct participants to respond to the research question. It is easy to allow creativity to seep in when developing a piece of visual inquiry and although creativity is not discouraged, it cannot be used to the detriment of not answering the question being asked.
- Arrange to have follow-up interviews with research participants to increase understanding of the collage/visual data. This would remove some of the risk associated with possible researcher bias.
- Develop a detailed framework for recording observations for each collage. This gives the researcher focus on what to look for in each collage.
- Develop a template that can be used to record the observations for each collage. The templates will facilitate analysis of all the observations.

There is a great opportunity to conduct further qualitative research on the difficulties of managing online and offline identities within educational settings where online learning is employed. Because of the rich text and image-based nature of online environments, future research should continue using both text and collage inquiry (or some other visual inquiry method) to capture the authenticity of student experiences of identity development within online and offline environments. This is needed to fully capture the emotional distress that emerges from the complexity of managing online and offline identities. It is also needed to piece together a better understanding of the many identity fragments (online and offline) that influence who individuals are, how they socialize, and how they learn, both individually and in groups. Turkle (1999) noted that "the many manifestations of multiplicity in our culture, including the adoption of online personae, are contributing to a general reconsideration of traditional, unitary notions of identity." It is obvious that a great deal of future research on offline and online identity will be needed to better understand how students develop and learn within educational institutions shaped by an increasingly digital world.

KEYWORDS

- **Online Social Networks**
- **Salience hierarchy**
- **Self-verification**
- **Text Findings**
- **Text-based data**
- **Visual research**

Chapter 11

Student 2.0 Revisited: The Paradox of Anonymity and Identity in the Digital World

Rocci Luppicini and Xue Lin

INTRODUCTION

Online learning is entrenched within the larger framework of the Internet and World-Wide-Web (WWW) where people interact each day, often under pseudonyms to protect private information and personal identity. A qualitative study is presented which explored the influence of online anonymity on student conceptions of self and of others. Sixty-six volunteer undergraduate students participated in online self-studies using narrative inquiry in response to the question, "How has the capacity for online anonymity influenced my conception of self and of others?" Students published personal narratives on a student researcher blog dedicated to collaborative self-study. Findings reveal that university students' self-concepts and conceptions of others are affected by the anonymity of cyber space and its mediating influences. Conflicting values of freedom and trust underlying students' positive and negative associations of online anonymity appeared to contribute to a schism in how university students' conceptualize themselves and others online. This gives rise to what the authors posit as the "paradox of online anonymity and identity". Implications for education are discussed and recommendations are made.

> *We judge on the basis of what somebody looks like, skin color, whether we think they're beautiful or not. That space on the Internet allows you to converse with somebody with none of those things involved.*
>
> —Bell Hooks

I could be anyone I might possibly want to be in the virtual world. Online anonymity provides the chance to set aside who we are and create who we want to be in the World Wide Web. The opportunity for deceit is rampant and has made me realize that everything online must be taken with a grain of salt.

— Student excerpt

Too often in educational research, scholars place the cart before the horse, in this case, the assumed success of online learning before the complex reality of the digital world people live in that does not necessarily lead to success. A good example of this, and the problematic within this chapter, is the topic of online anonymity. What then, is anonymity? Pfitzmann and Hansen (2008) defined anonymity as "the state of being not identifiable within a set of subjects, the anonymity set." Communication on the Internet is often described as "anonymous." Multiple levels of anonymity (visual anonymity, dissociation of real and online identities, and lack of identifiability) are thought to have affects on different components of interpersonal motivation (Morio & Buchholz, 2009), which can influence one's self-conception. Morio and Buchholz (2009) suggested that by creating a new online identity an individual can virtually become a different person, with different personalities and values. Some individuals even adopt a new gender, age, and/or a race. This is because online anonymity allows there to be (in varying degrees) a dissociation of real and online identities where someone's online behaviors are not directly tied to that person's physical self. Because of this dissociation, there tends to be less accountability for online behaviors compared to behaviors in "real" life (Morio & Buchholz, 2009). This leads to one of the biggest conundrums associated with the social use of modern day Internet.

On the one hand, an anonymous online identity can be empowering because it allows individuals to embrace their diversity, engage in identity exploration, and occupying social positions which may be difficult to occupy in real life. Turkle (1996) noted that the Internet allows individuals to hide parts of their identity which might lead to discrimination (e.g., race or gender). And as stated by Gies (2008), "anonymous online settings are empowering because they facilitate identity exploration, or occupying identity positions which may be difficult to occupy in real life". On the other hand, a number of scholars have pointed out that online anonymity is problematic and can lead to online flaming, cyber-bullying, information theft, and defamation. These contradictory consequences of anonymity warrant serious consideration for the design of e-learning.

This study is based on two assumptions: (1) that online learning environments are a microcosm of the larger macrocosm of online environments that permeate the digital world of the Internet and WWW, and (2) that meaningful collaboration (and collaborative learning) depends on trust in how individuals present themselves and interact with one another. As a growing part of the social world and educational system, the Internet and WWW are becoming more and more entrenched within the lives of students within and beyond the walls of the educational institution. However, this digital world is unlike the natural world in that individuals have greater control over how much of themselves they reveal to others with the option of remaining completely anonymous, distorting personal information about themselves, or creating alternative online identities that do not connect with who they are in real life. How does online anonymity influence student conceptions of themselves and others? Does anonymity affect trust required for teamwork and community building? To what extent will these conceptions affect student motivation to engage in meaningful online interaction? And to what extent will these conceptions influence students' online actions? These are some of the questions that lead to following research.

One popular view on online identity is based on the notion of self-concept as an expression of self-cognition and self-expectancy within anonymous computer-mediated environments. Gonzales and Hancock (2008) used a public commitment framework, Self-Presentation Theory, to examine how computer-mediated self-presentations can alter identities. Their findings suggest that the idealized versions of the self presented online may reinforce "actual" self-perceptions unrelated to the mediated interactions.

In other words, not only can people take advantage of online anonymity to explore new aspects of the self (Bargh, McKenna, & Fitzsimons, 2002), but they also can take advantage of the public nature of the Internet to help realize idealized concepts of self. This has important implications within the educational context where online learning is often used as a supplement to traditional classroom learning to provide opportunities for socialization and collaborative work not possible within large classes. This is particularly true of undergraduate university classes where 100–1000 students who do not know each other may use the online component to get acquainted before interacting in class where the idealization of online self-concepts becomes realized.

The purpose of this study is to uncover influences of online anonymity on university students' conceptions of self and others. A phenomenological study of a class self-study blog assignment was conducted drawing on existing online identity research and collage inquiry techniques aligned with phenomenological research (Butler, 2008). The guiding research questions were as follows: How has the capacity for online anonymity influenced university students' conceptions of self and of others?

METHODOLOGY

This study draws on a phenomenological approach to examine how the capacity for online anonymity has influenced university students' conceptions of self and of others. According to Creswell (2009), phenomenological research is a strategy of inquiry in which the researcher describes the meaning of individuals' lived experiences of a particular phenomenon, refrains from importing personal insights and familiarity with the phenomenon, and ultimately seeks to identify the overall "essence" of such experiences. Therefore, as employed to this study, phenomenological research method is ideal for seeking a deeper understanding of identity shaping when people experience a particular phenomenon like online anonymity.

This study was a classroom research project designed to explore a class self-study blog created as part of an undergraduate course in qualitative research. Students were asked to create short personal narratives in response to the following question: "How has the capacity for online anonymity influenced my conception of self and of others?" The blog allowed students to design and anonymously publish personal narratives about online anonymity and their identities. Participation in the study was voluntary and students were in control of what they contributed to the class blog.

In terms of data analysis, narratives from the blog were analyzed using the following steps. First, all transcriptions on the blog were examined to create a list of significant themes which illustrated common aspects of the phenomenon that describe what was experienced (horizontalization). The themes addressed advantages and disadvantages of online anonymity as well as statements about online identity. These significant statements were then used to develop clusters of meanings (meaning units) which reflected common themes of online experience among participants (Moustakas, 1994). Based on the themes and meaning clusters identified, descriptions were juxtaposed to explore commonalities and get at the "essence" underpinning participants' experience of how online anonymity influences university students' identity (Creswell, 2007).

FINDINGS

The analysis of the student self-study blog narratives revealed a variety of perceived influences of online anonymity on participants' self-conceptions and conceptions of others. First, it was discovered that many of the participating university students had both positive and negative associations with online anonymity. Key themes, clusters of meaning, and frequencies are presented in Table 11.1 and discussed below:

Table 11.1. Breakdown of the self-study narrative analysis.

Themes	Clusters	Frequency
Opportunity	Positive associations	9
Confidence		6
Freedom		23
Happiness		3
Benefit for learning		8
Fearlessness		7
Identity development opportunity		10
Sharing		5
Openness		6
Creation		3
Easiness to change		3
Togetherness & connectivity		6
Trust		20
Fakeness	Negative associations	10
Caution/danger		29
Hiding		21
Irresponsibility		8
Distance		3
Privacy		7
Illegal activities		3
Vulnerability		2
Untrustworthiness		11
Irretrievability		4
Identity		24
Communication	Neutral associations	15
Technology		11
Anonymity		28

Positive Associations

As indicated in Table 11.1, university students had many positive associations with how online anonymity influenced their identity. The themes are: freedom, trust identity development opportunity, confidence, happiness, benefit for learning, fearlessness, independence, sharing, openness, creation, easiness to change, togetherness and connectivity, and trust. The most predominant theme for students was anonymity and freedom. One student commented that "online anonymity has created a sense of freedom that can be invigorating" (student excerpt). Other comments from other students elaborated on the meaning of freedom provided by the anonymity of the Internet:

- "There are really no barriers to what you can and cannot say" (student excerpt).
- "I feel to ask whatever questions I have without potential repercussions or embarrassment" (student excerpt).
- "There is a sense of freedom online that you cannot find anywhere else, you don't know who you are interacting with and neither do they" (student excerpt).
- "Because of online anonymity, people may be able to express themselves more freely" (student excerpt).

The second most common theme touched on by students concerned the connection of online anonymity to trust. One student commented, "I personally prefer face-to-face interactions and have no trust in the Internet. The vagueness of individuals online has made me weary of what I say to whom, and has made my online preferences extremely private and secure" (student excerpt). Another student draws a similar comparison in how "Even though technology is making communication more efficient with each passing day, a face-to-face conversation is still one of the warmest, most sincere experiences one can have in a relationship in order to build solid trust" (student excerpt). These comments get at the perceived difference between trust within their online and offline experiences. Other comments focused more on trust online:

The fact that it is so easy now for sources to impact the world of media either with a fake identity, or even without one at all, has decreased my trust in many forms of online media, including daily news and even some forms of communication with acquaintances.

The capacity for online anonymity has definitely influenced my level of trust. It has caused me to question the truthfulness of others. (student excerpt)

In my perception of others, Internet anonymity has made me highly skeptical and distrustful of the actions of others online. I feel like people are untrustworthy when left with such an open medium. I fear that illegal activities are going to become more pervasive without increased control. (student excerpt)

Online, however, I don't trust anyone or anything. Often when people can be anonymous they say and do things they otherwise would not. (student excerpt)

The third most common theme touched on by students concerned the connection of online anonymity to identity experimentation and development. One student stated, "I see others as windows of opportunity to things and knowledge that I have the great ability to access and discover" (student excerpt). As summed up by one student. "You can be anybody you want" (student excerpt). This tied into other themes revolving around identity development and preservation of oneself. As stated by on student, "It [anonymity] can give you a chance to freely communicate with strangers with minimal risk to self" (student excerpt). What most of the positive associations had in common seemed to link to a sense of personal control and freedom provided by online anonymity that allowed students to express themselves, seek information, communicate, and engage in identity development by experimenting with their online personae.

Negative Associations

Table 11.1 also highlights how university students had many negative associations with how online anonymity influenced their identity. Key negative influences of online anonymity found in student narratives included: danger, hiding, suspicion, untrustworthiness, fakeness, irresponsibility, privacy, irretrievability, distance, illegal activities, and vulnerability. Student attitudes toward online anonymity were often tied to perceptions of untrustworthiness. One participant said he/she did not trust anyone or anything online, "The fact that it is so easy now for sources to impact the world of media either with a fake identity, or even without one at all, has decreased my trust in many forms of online media, including daily news and even some forms of communication with acquaintances" (student excerpt). Many students believed that Internet anonymity has made them highly skeptical and distrustful of the actions of others online. As stated by one student, "in my perception of others, Internet anonymity has made me highly skeptical and distrustful of the actions of others online. I feel like people are untrustworthy when left with such an open medium" (student excerpt). Findings indicated that student question online information and people they come across. Moreover, participants highlighted their worries and sense of vulnerability because of dangers online anonymity brings. As one student pointed out "by escaping identity you escape responsibility, therefore all actions made on behalf of online anonymous Internet users are made without the fear of consequences" (student excerpt). Many online dangers were highlight by students including, financial insecurity on the Internet, identity theft, online deception, online dating risks, and acts of online hate. These were some of the reasons provided by students who decided to hide their true identities to protect themselves from danger. These findings seem to indicate that university students' negative associations with online anonymity are centered on untrustworthiness of online information which leads to feelings of uncertainty, fear, and danger as reflected in student statements concerning fake information on the Internet, invisible distance between users, widespread illegal activities, and irresponsible behavior. The next section expands on the stated findings by offering an interpretation to further explain students how students perceive the connection of online anonymity to how they view themselves and others.

INTERPRETATION

Above mentioned findings allude to the fact that people are born into a particular time period and participate in specific social situations that provide them with numerous resources for the construction of conceptions of self and of others. Within this digital world, the resource is the Internet and many of the social situations that shape the construction of identity take place within the Internet arena. Through this phenomenological study, the role of how online anonymity influences university students' construction of self-conception is highlighted. More specifically, findings suggest that: (1) online anonymity is perceived by individuals as a double-edged sword that brings both positive and negative influences into their lives; (2) individuals associate online anonymity as a positive influence on their personal identity development because it provides freedom and opportunities; and (3) individuals associate online anonymity as a negative influence in their conception of others because it makes them suspicious of information and people encountered on the Internet. These three findings are interwoven in interesting ways that give rise to the paradox of online anonymity which provides opportunities for greater understanding and development of one's self-concept while hindering the development of understanding and trust in others' presented self-concepts. That is, online anonymity provides individuals a high level of personal control and confidence in their self-concepts but a low level of personal control and confidence concerning others presented selves. The implications of this are discussed below.

Online Anonymity: A Double-Edged Sword

Findings from this phenomenological study that portray online anonymity as a double-edged sword are validated by other research. On the one hand, there are clear benefits to be gained from online anonymity. It promises users freedom and opportunities. Certainly the lack of user knowledge about another's race, gender, appearance, and socioeconomic background can has been found to facilitate freedom of thought and expression by promising people a possibility to express opinions that, for fear of reprisal or ridicule, they would not or dare do otherwise (Moral-Toranzo, Canto-Ortiz, & Gomez-Jacinto, 2007; Nissenbaum, 1999). Parks and Floyd (1996) discussed that online anonymity allows people the opportunity to acquire social skills, to conquer introversion, and "to transcend the limitations they experience in face-to-face settings" (p. 83). For instance, Froomkin (1996) investigated online anonymity in the context of free speech and democracy, arguing that online anonymity may be the best option for ordinary individuals to protect themselves from governments' and private corporations' active use and profiling of their personal information in the networked environment. In a similar vein, Nissenbaum (1999) argues:

> Anonymity may enable people to reach out for help, especially for socially stigmatized problems like domestic violence, HIV or other sexually transmitted infection, emotional problems, or suicidal thoughts. It offers the possibility of a protective cloak for children, enabling them to engage in Internet communication without fear of social predation or perhaps less ominous but nevertheless unwanted overtures from commercial marketers. Anonymity may also provide respite to adults from commercial and other solicitations. It supports socially valuable institutions like peer review, whistle-blowing, and voting. (p. 142)

Conversely, there is a large literature addressing the consequences of exposure to online anonymity. Research has noted how anonymity increases disinhibited, hostile behavior, and extreme decision making and reduces attraction within computer-mediated groups. These negative aspects of online anonymity have been linked to numerous factors discussed in the research literature including: lack of interpersonal cues, feelings of deindividuation, reduced evaluation concern, and a more impersonal and task-oriented attentional focus (Kiesler, Siegel, & McGuire, 1984; Walther, 1992).

The duality of online anonymity is revealed in student responses where there is a recognition of positive and negative aspects of online anonymity. This is apparent in a number of student statements where online anonymity was, at the same time, portrayed in a positive and negative light:

> The Internet for me has become a personality building and accentuating tool, it helps me know more about myself and others, defining who I am. *Yet*, in my perception of others, Internet anonymity has made me highly skeptical and distrustful of the actions of others online.

> I think many people are concerned about what they share online and their privacy, *but at the same time,* they still want the world to know what they're thinking and doing.

> Online anonymity has made me both more skeptical and cautious of others *but at the same time* open and more confident about myself when communicating online

> The positive aspects of online anonymity I see for myself and others are that Websites, forums, and blogs are places that allow people to ask questions they may be afraid to in real life. This is a double-edged sword, however, because people can attack others without fear because their identity is hidden.

Anonymous Internet: A Safe Space for Self-Concept Development

Identity is an important part of the self-concept. Self-concept is the totality of a person's thoughts and feelings in reference to oneself as an object (Rosenberg, 1986), and identity is that part of the self by which individuals are known to others. Turkle (1996) argued that in anonymous Multi-User Domains (MUDs), people can disguise aspects of identity which might lead to discrimination, such as race or gender, and so can perform a range of identity positions, hiding marginal identities, and becoming part of the mainstream. So to speak, online anonymity promises people confidence, which is something they may be lack of in real life. Vieta (2004) investigated the relation between online fantasy and anonymity in relation to offline life. Online fantasy and anonymity could be thought of as an instrumental and temporary decision by the interactant to suspend the "external world" and engage purely in fantasy and play (Vieta, 2004). So without revealing real identity, online anonymity helps people to escape from real life and reconstruct their virtual identities in line with their expectancy. In a study of *Facebook*, Zhao, Grasmuck, and Martin (2008) argued that *Facebook* enables the users to present themselves in ways that can reasonably bypass physical "gating obstacles" and create the hope for possible selves they are unable to establish in the offline world. Such "digital selves" are real, and they can serve to enhance the users' overall self-image and identity claims and quite possibly increase their chances to connect in the offline world. Implicit in this claim is the assumption that anonymity in cyberspace is potentially empowering because it facilitates identity exploration, by allowing individuals to test out various identity positions which may be difficult to occupy in real life (Nissenbaum, 1999).

Anonymous Internet: A Dangerous Ground for Understanding and Trusting Others

Because people are really unknown to others online, mistrust is a natural reaction in the face of unfamiliarity and uncertainty. And while virtual environments are valuable as places where individuals can explore their inner diversity with others, there is still the need for an authentic experience of others that is lost with online anonymity. One student expressed that "everyone is a shadow represented online by only a screen name" (student excerpt). The fear of online anonymity is the fear that in the culture of simulation, a word like authenticity can no longer apply and the true nature of individuals is forever hidden (Turkle, 1996, p. 254). This was mirrored in this study by participants expressing their worries about anonymous others online. As indicated by one student, "the capability of interacting online without showing one's true nature leads me to believe the majority of individuals are afraid to reveal their complete genuine identity" (student excerpt). What compounds this problem of trust and fear is that fact that individuals often exchange more personal information about themselves online than offline which makes it difficult to ignore. According to Pauwels (2005), "People may reveal more of their identities and provide more valid information than in face-to-face situations, where their anonymity cannot be secured and where their conduct may be challenged directly" (p. 605). Due to fear, suspicion, and lack of trust within the anonymous online realm, it is easy to see how online anonymity can negatively influence the development of individuals' conceptions of others'.

Paradox of Online Anonymity and Identity

Drawing together key findings and interpretations alludes to a complex picture of online anonymity and self-concept marked by core values affected by key dimensions of online anonymity, which in turn, affect how individuals approach their own self-concepts and those presented by others online. The conflict of core values with key dimensions of online anonymity constitutes the paradox of online anonymity and identity which affects the development of self-concept and conceptions of others. Table 11.2 presents a breakdown of this paradox discussed below.

Table 11.2. Breakdown of the paradox of online anonymity and identity.

Core Value	Dimenion of Online Anonymity	Self-Concept	Conception of Others
Freedom	Control	High degree of freedom and control in shaping online interactions and self-presentation.	Low degree of freedom and control in how others present themselves and interact online.
Trust	Authenticity	Perceptions of self as a multiplicity of authentic selves in development.	Peceptions of others multiple representations of self as untrustworthy and insincere.

Table 11.2 presents a model highlighting a combination of core values and dimensions of online anonymity which influence how individuals approach their online self-concept and conceptualizations of others online. Freedom is a core human value which

is affected in different ways within anonymous social relations online. With respect one's own self-concept, individuals have a high degree of freedom and control in shaping their online interactions and self-presentation. Woo (2006) states," The degree to which a person has autonomy regarding the control of information about themselves and their relationship to outside information, is an important factor influencing the establishment of the person's identity" (p. 957). As indicated by one student in this study when describing her view of online anonymity, "I feel powerful and in control with the ability to do whatever I want and get whatever I need" (student excerpt). At the same time, individuals have a low degree of freedom and control in how others present themselves and interact online. One would expect that individuals will invest more time and effort in their own online self-concept development compared to their investment in trying to control how others present themselves online.

Trust is a core human value which is affected in complex ways within anonymous social relations online. With respect one's own self-concept, individuals tend to perceive themselves as a multiplicity of authentic selves in development. As discussed elsewhere in the chapter, researchers have found that online anonymity provides an ideal social space for experimenting with and advancing multiples aspects of oneself that individuals would not do offline. And because individuals are aware of their actions, whatever individuals do online is authentic in the sense of being an experience they had. This, however does not apply to individuals conceptions of others which cannot be experienced first hand, but only observed or inferred. In the offline world (real world), appearances and behaviors of others cannot be directed experienced but they can be observed to obtain at least some level of certainty in the authenticity of their self-presentations and actions. This is not the case within an anonymous online environment where the authenticity of another's experiences must be based on indirect observation and inferrances that take place within a mediated environment. There is simply less certainty and greater room for deception within anonymous online environments which limits trust in individuals perceptions of others. As stated by one student:

> I am influenced by the studies and the news media that report on the deception—trickery–that takes place online because of online anonymity. Online anonymity creates opportunities for criminality. For this reason, I choose to be cautious online: there are no quick solutions, a 'friend' is not a friend because the screen message says so, you are not special because Website X says you are, words are not necessarily meaningful. (student excerpt)

Others' multiple representations of self within anonymous online environments may be perceived as untrustworthy and insincere for that reason. One would expect that individuals will be more invested in their own online self-concept development compared to their investment in the how others present themselves online. Overall, the model helps explain why individuals are more invested in their own online self-concept development compared to their investment in the how others present themselves online. This has significance for the advancement of education and research which will be touched on in the discussion below.

DISCUSSION

Significance of Study and Future Focus

As a growing part of the social world and educational system, the Internet and WWW are becoming more and more entrenched in the lives of students within and beyond the walls of the educational institution. However, this digital world is unlike the natural world in that individuals have greater control over how much of themselves they reveal to others with the option of remaining completely anonymous, distorting personal information about themselves, or creating alternative online identities that do not connect with who they are in real life. This chapter uncovered how core human values (freedom and trust) are influenced by key aspects of anonymous online environments, which in turn, shape how individuals present themselves and identify with one another. This is important for education because students' conceptions of self and others affect social behavior and how they approach learning and group work within online learning environments.

Implications for E-Learning Design and Online Instruction

There are a number of implications for e-learning design and online instruction. First, the use of online group work within anonymous online environments can be used to leverage student motivation and structure self-directed learning. As expressed by one student, "My ability to be anonymous online has enabled me to feel free to surf the maze of the Internet, looking up things that I want to learn more about, learning more about myself and my interests" (student excerpt). Online learning designers and instructors should consider integrating anonymous online group work into their online instructor to help stimulate student personal growth and self-directed learning. This is particularly well aligned with constructivist oriented pedagogical methods. Second, anonymous online learning environments can be used by online learning designers and instructors to leverage diversity training and multicultural education. As stated by one student, "Since most people have a willingness to share more online, this shows that we ultimately live in a world that is connected together more than ever and we should take advantage of this by teaching and learning from others around the world" (student excerpt). Online learning designers and instructors can take advantage of the freedom of expression offered to students by anonymous interaction to help them explore delicate issues that challenge their personal sense of self and views of others. This could be especially useful within moral education where there is desire to raise multicultural awareness and acceptance of others.

Perhaps the greatest challenge for educators in integrating online anonymity into e-learning design and online instruction revolves around trust. Educators must develop strict guidelines for monitoring (policing) any educational application of online anonymity to guard against the dangers discussed earlier in this chapter and minimize the threats to student trust that accompanies anonymous online interaction within the larger Internet and WWW. Educators must integrate media and Internet awareness instruction into the curriculum to ensure that students are aware of the dangers and risks when engaged in anonymous online learning. There is a promising opportunity for an educational re-tooling of the Internet and WWW for learning purposes that is worth exploring, but only with the proper safeguards in place.

KEYWORDS

- **Multi-User Domains**
- **Online Instruction**
- **Personal identity**
- **Private information**
- **Self-Concept Development**
- **World-Wide-Web**

Chapter 12

Education for a Digital World: Current Challenges and Future Possibilities

Rocci Luppicini

INTRODUCTION

How can diverse educational technology needs across the various departments within a university be met? How can the faculty use distance education as an added value to augment the regular on-campus course offerings? Will there be adequate faculty member training opportunities for those interested in integrating new technologies in the classroom? What online course management system is best suited for the faculty? What faculty member and student input is available or can be gathered to find out about needs and interests regarding the use of technology in teaching and learning? These are some of the questions that emerge during large-scale technology planning initiatives. This chapter provides an expert reflection on one institution's response to technology planning taking place at the faculty level. The author identifies an underlying barrier that can block university planning strategies, referred to as, *the paradox of academia*. It explores the dangerous separation between macro level (university level) and micro (faculty level) planning issues that needs to be made to better focus planning efforts on core issues at the appropriate level of application. The chapter draws on existing scholarship on Technoethics to shed additional light on the challenge of technology planning at the university level. Based on insights drawn, suggestions are made on future technology planning strategies.

This edited volume comes at a pivotal time for many professionals working within educational institutions where technology is at the core of institutional planning for the future. Like many other educational institutions around the world, my own university faculty recently launched an institution wide planning campaign (called Vision 2015) to define our main priorities over the next 10 years and how to get there. Not surprisingly, technology was one of the key areas within our faculty and I was part of the advisory group responsible for identifying the priorities for technology planning and integration intended to serve the myriad of needs and interests within this academic unit.

Anyone that has been involved in this type of technology planning process has first hand experience with how daunting and complex this task is. There is no easy solution and no way to satisfy all interests concerning every aspect of the move forward. What is essential is to devise an approach to planning that takes the needs and interests of all stakeholder groups into consideration so that decision-making reflects the general vision that can be agreed to and worked toward as an educational community.

In the discussion that follows, one institution's response to technology planning is reflected on to gleam practical insights into technology planning strategies and challenges (or stratagems and spoils to borrow from Bailey's 1969) seminal work on political leadership and decision-making. Based on insights drawn from this institutional technology planning campaign, suggestions are made on future technology planning strategies under the framework of Technoethics.

STRATEGIC PLANNING AND THE CHALLENGE OF SHARED VISION

The University of Ottawa (henceforth called the University) is located in the nations capital city and is North America's premier bilingual university. It is a large urban comprehensive university (undergraduate and graduate) with over 38,000 students, offering over 360 undergraduate and 110 graduate programs across 10 faculties. In 2010, the University embarked on a strategic planning initiative to help define key priorities over a 10 year period. As stated by the University President Allan Rock:

> Strategic planning can generate major change, re-align the institution and bring key challenges and issues into brighter, sharper focus. This exercise is a community-wide effort. Everyone, including our external partners, is called upon to ask fundamental questions about the nature and mission of the institution, to reflect on the conditions for success and, ultimately, to produce a widely supported strategic plan that serves as a call to action throughout the institution. Strategic planning helps craft a collective vision, which in turn produces a reassuring measure of stability and continuity within our community. It also provides individual units with the opportunity to define themselves as part of the institution as a whole and to bring their actions into step with a common vision. (p. 1)

In order to reflect on core issues roundtables were created to address the following five proposed goals: (1) to play a leadership role in promoting Canada's official languages. (2) to achieve prominence on the international stage, (3) to create knowledge, to discover and to invent, (4) to offer students an unparalleled university experience, and (5) to care for our community. Roundtable consisted of approximately 15 people that represented the University community and the external community (University of Ottawa Vision 2020, 2010).

Within the Faculty of Arts (henceforth, the faculty), a five year vision plan is currently being initiated and roundtables, consisting of professors, support staff, and students, have been created to address the following five areas of concentration: (1) fundamentals in education, (2) program delivery mode (technology), (3) undergraduate student experience, (4) graduate student experience, and (5) research support. The roundtable focused on program delivery mode deals with core technology (and other) considerations for campus and distance teaching and learning. This committee met numerous times to discuss challenges, needs, and interests from across the faculty with regard to technology planning and program delivery goals for 2015. Committee meetings focused on identifying core faculty interests and needs regarding program delivery mode and technology integration. The objective of the meetings was to create a list of key questions to address within a public forum open to all members of the faculty (the faculty version of a community town hall meeting).

The problem emerging from this initiative revolves around trying to align faculty level needs and interests onto the needs and interests of the university as a whole. In this case, a misalignment of faculty and university level focus is apparent in their respective areas of roundtable concentration as depicted in Table 12.1.

Table 12.1. University and faculty vision areas of concentration.

University Vision 2020	Faculty Vision 2015
Leadership role in promoting Canada's official languages	Fundamentals in education
Prominence on the international stage	Program delivery mode (technology)
Knowledge creation	Undergraduate student experience
Offer students an unparalleled university experience	Graduate student experience
Care for our community	Research support

Attempting to map the faculty areas of consultation onto the university consultation areas proved difficult. One could infer that the university level concern for offering students an unparalleled university experience partially maps onto three of the five faculty areas, namely: fundamentals in education, undergraduate student experience, and graduate student experience. In addition, there is obvious overlap between knowledge creation and research support (which provides a basis for knowledge creation). However, it is unclear how the university focus on community outreach and bilingualism maps onto the faculty focus. Similarly, the faulty focus on technology and program delivery does not appear anywhere in the listed university consultation areas or within the Vision 2020 documentation (University of Ottawa Vision 2020, 2010).

There are two important points to be observed: First, there is a misalignment of educational planning scope which is a five year plan at the faculty level but a 10 year plan at the university level. As stated within the university of Ottawa Vision 2020:

> Since we believe that reflecting on our institution's nature every five years is not necessary and that a major change in direction would prove difficult, our aim for the next strategic plan is to adopt a vision that takes us right to the year 2020. Though the vision does indeed extend into the very long term, the initial plan of action will cover three to five years, given how hard it is to predict potential changes. (p. 2)

Second, there also appeared to be a misalignment between the university planning strategy and that of the faculty concerning program delivery mode and technology planning issues. While technology concerns were apparent at the faculty level, they were not represented within the university level strategic documentation. The consequence of organizational misalignment between the parts and the whole, can lead to the formulation of faculty level action plans that do not align with organizational objectives of the university as a whole. This misalignment can be observed by examining the key issues focused on at the faculty level which affect the university as a whole. Key questions explored at the faculty level included the following: How can diverse educational technology needs across the various departments be met? How can the faculty use distance education as an added value to augment the regular on-campus course offerings? Will there be adequate faculty member training opportunities for those interested in integrating new technologies in the classroom? What online course management system is best suited for the faculty? What faculty member and student input is available or can be gathered to find out about needs and interests regarding the use of technology in teaching and learning?

What faculty member and student input is available or can be gathered to find out about student needs and interests regarding the use of technology in teaching and learning? One of the first things our Vision 2015 faculty committee did was consult outsider research and existing faculty level student surveys carried out by one of service units within our faculty. This provided a partial snapshot of student experiences, needs, and interests concerning technology use within the classroom. This was a fruitful activity which helped situate and guide the planning committee in some areas. It provided a base to begin conversation about what faculty needs will be based on what they currently are and what is desired.

How can diverse educational technology needs across the various departments be met? The answer to this question would provide a view of only some departmental needs within the university. Unfortunately, this does not provide an accurate snapshot of the diversity of university level needs. Without the coordination of a similar inquiry from all other units within the university, the potential value of this inquiry is not fully realized. And in fact, it may lead to intra-university conflict. Part of the reason is that large-scale technology infrastructure costs are typically made at the university level and university level purchases of technology products (software, computers, etc.) are best done at the university level due to the economy of scale savings leveraged through university level purchasing.

How can the faculty use distance education as an added value to augment the regular on-campus course offerings? Even at traditional campus-based universities like ours, the drive for distance education courses to extend course offering potential is attractive. As a typical large urban university which prides itself on openness, there are many part-time students and co-op programs interested the flexibility of distance education. This is particularly attractive for students living outside the region and students wishing to take summer courses which accommodate daytime jobs and work study program schedules. However, because distance education can impact the university reputation, branding, and faculty member copyright over course materials, discussing distance education at the university level should occur before individual faculties enter the game. Today, the distance education market is huge and distance education providers, including some academic publishing companies, are all too ready to provide client assistance for hosting distance education program delivery complete with e-books, other e-resources, testing materials, and textbooks. It is tempting for faculties to fall into the trap of letting private companies take over and provide this one-stop shopping experience for courses for assigned part-time lecturers to free up full-time faculty member time for research and service work. This can often lead to faculty member uprising over course delivery quality, intellectual propriety disputes, and in some cases, union uprising. Carefully negotiated agreements with third party providers need to be worked out using university legal services, union representatives, faculty member, and student input for more informed organizational decision-making to proceed.

Will there be adequate faculty member training opportunities for those interested in integrating new technologies in the classroom? Technology integration required proper marketing and technical support, which is costly. Implementing new technologies is time and resource demanding. Faculty members can get entrenched within one type of course delivery mode and not want to change their habits, particularly if past technology implementation efforts have failed. For many faculty members, teaching is all about trust and getting them to invest their limited time to learn a new system where failure has occurred in the past is asking a lot. There are also situations where departments have recognized this and used their own operating budgets to provide additional technical support. The problem with this is that this can take away from other operational areas. I once worked within a university department that used their budget in this way and it resulted in technology rejection when the university required that faculties cut departmental spending of non-essential spending. Because additional technical support funding was not part of the regular operational budget, the department had to cut the technical support services which resulted in technology failure and user (faculty member and student) frustration.

What online course management system is best suited for the faculty? Online learning management systems are a major part of university teaching and learning at many universities but deciding which system to use is challenging since so many are available (ANGEL LMS, Blackboard, CCNet, Chamilo, Claroline, Dokeos, DoceboLMS, ILIAS, Moodle, Pass-port, OLAT, SharePointLMST, etc.). This is another complex issue for faculties to navigate. Because the cost of implementing and maintaining such systems is expensive, getting university level "buy in" allows the university to absorb this burden, often at a lower overall cost if multiple faculties adapt the same system.

What becomes apparent in the above mentioned areas of discussion is that there is separation between macro level (university level) and micro (faculty level) planning issues that needs to be made to better focus planning efforts on core issues at the appropriate level of description. Based on a social systems view, faculties are parts of and constitute the university as a whole. The challenge of technology planning exercises like this are to figure out how the faculty structures and processes interrelate with the university as a whole. The next section offers a social systems view to shed additional light on the challenge of technology planning.

RE-ALIGNING UNIVERSITY PLANNING VISION

Through experiences gathered in the academic trenches, I have come to discover what I call the *paradox of academia* that seems to plague organizational planning processes within educational institutions I have visited and worked at over the years. It is a paradox of professional identity leading to overly narrow view of what matters and to whom. On the one hand, faculty members are academic experts in their respective fields working with colleagues to carve out various niche areas of academic knowledge. On the other hand, faculty members are also core members of their university community defined by teaching activity, service work, and research carried out within a particular university which has its own core values and goals as an academic institution. It is the conflict in professional identity (professional identity as an academic leader vs. professional identity as an university member) that often leads to conflict when it pertains to university decision-making. This identity conflict arises from the conflation of social roles and the conflation of the notion of knowledge and information.

The *conflation of social roles* of academics leads to the view of the academic as the lone wolf in academia independent their academic institution. A colleague once told me as a young academic that academia was NOT a team sport. This is not surprising since many academics must drift from one educational institution to the next to secure tenure track and tenured positions. It is hard to develop a sense of loyalty and institutional identity within this professional socialization situation. It is a common criticism that academia operates by separate disciplines and fields creating silos to build up their domains (and the reputations of contributing scholars along with it). It is not a problem that academics are "experts" in specific areas and contributors to these areas. It only becomes a problem when faculty members become conceptually fixed within their disciplinary orientation situated within a particular department or faculty. This runs the risk of creating group think that prevents the realization of better options available and in-group biases that leads to an overestimation of one groups own place and importance within the university. After all, why should a faculty member from medicine care about what is going on within a communication department? It is this narrow mindset that prevents many fruitful interdisciplinary exchanges in overlapping areas of expertise (i.e., health communication). However, without coming to terms with the social role of academics as members of a large university community, it is difficult to imagine how coordinated university level planning activities could succeed.

The *conflation of information and knowledge* is a second major problem within organizational planning in that it takes removed the person from the (see Technoethics). Luppicini (2010) distinguishes how the metaphor of information deals with information or data whereas knowledge depicts a human activity and is more closely aligned with organized aspects of human life and society. When academics engage in knowledge production, they are not just creating information from a boundless ivory tower for consumption. Rather, they are expert knowledge workers (researchers, instructors, consultants, etc) participating in organizational practices that place the organization (the university) at the forefront of their professional activity within society and the contemporary economy. As expressed by Dyson, Gilder, Keyworth, and Toffler (1994) in describing organizational work within the current knowledge-based economy, "In a Third Wave economy, the central resource—a single word broadly encompassing data, information, images, symbol culture, ideology, and values—is actionable knowledge". And because knowledge is connected to human activity, it is a value-laden and connected to the organization (university) in which it evolves. Under this metaphor of knowledge as a human activity, understanding knowledge and knowledge production requires an understanding and appreciation of the organization and how knowledge production is socially constructed within it (Pinch & Bijker, 1987). Although faculty members all have areas of expertise and professional affiliations, they all come together within the university as a whole to help define it. This is how and where it all happens—It is the employer and site of professional activity within society that defines and is defined by the organizational members that constitute it (students, staff, administration, faculty members) and member activities (teaching, learning, research). In other words, faculty members are not just information producers at large in society, but are professional knowledge workers and contributing members to the university, a social system where a variety of other organizational practices and perspectives must be respected and accommodated.

The narrowness of mindset resulting from these two conflations can create organizational barriers to coordinated action that requires the participation from all areas of the university organization. It is this mindset that can prevent one area of the university from understanding what is going on in the other parts. And when it comes to organizational level technology planning, it is core that members from each part of the university understand the perspectives of members from the other parts and how this shapes (and is shaped by) the university as a whole. This is basis of social systems thinking and grounding for the following technology planning solution described here.

TECHNOETHICS AND TECHNOLOGY PLANNING WITHIN EDUCATIONAL INSTITUTIONS

This section offers a sketch of Technoethics, an approach to technology planning from a social systems perspective of technology and human value. According to the *Handbook of Research on Technoethics*:

> Technoethics is defined as an interdisciplinary field concerned with all ethical aspects of technology within a society shaped by technology. It deals with human processes and practices connected to technology which are becoming embedded within social, political, and moral spheres of life. It also examines social policies and interventions occurring in response to issues generated by technology development and use. This includes critical debates on the responsible use of technology for advancing human interests in society. To this end, it attempts to provide conceptual grounding to clarify the role of technology in relation to those affected by it and to help guide ethical problem-solving and decision making in areas of activity that rely on technology. (Luppicini, 2008, p. 4)

Under this umbrella, the university is viewed as a social system characterized by human values embedded within knowledge production and technology planning activities. Technoethics acknowledges the value-laden nature of knowledge production within organizations and the need to connect the strands of knowledge created within the university to inform technology decision-making at the organizational level. As noted by Luppicini (2010):

> The advent of technology in many areas of human activity has given rise to a plethora of technology focused programs of ethical inquiry scattered across multiple disciplines and fields. Efforts to reach an understanding of ethical aspects of the various types of technology are challenged by the tendencies within academia to create silos of information in separate fields and disciplines. Technoethics helps connect separate knowledge bases around a common theme—technology.

How does Technoethics apply to this particular case? As indicated above, the first step to faculty level technology planning is to figure out what are the macro level (university level) and micro (faculty level) planning issues. This requires a deep understanding of the organization and any challenges that can block successful macro level technology planning (i.e., paradox of academia). It is important in organizational technology planning to take a step back to look at the university as a whole constituted by multiple stakeholders with diverse values, interests, and needs. This chapter assumes that appropriate technology planning at the organizational level requires a mutual understanding of the multiple perspectives within the university where organizational decisions are formulated and applied. As discussed above, the conflation of multiple social roles of academics and the conflation of the concepts of information and knowledge represent traps that must be avoided.

The next step is to identify strategies and rules to govern technology planning implementation. At the university level, the technology planning is about developing effective rules and strategies based on the values and knowledge of its university members. The question is, what types of rules or strategies are available to guide organizational technology planning. Because organizations are made up of people with varying interests and values, decision-making has political aspects. For instance, Bailey's (1969) *Stratagems And Spoils: A Social Anthropology Of Politics* provides a practical view of political decision-making processes within organization. The author likens such processes to tactical games involving players, teams, a playing field, an audience (fans), goals, rules, and referees. Bailey (1969) makes an important distinction in his work between normative rules and practical rules:

> Normative rules are very general guides to conduct; they are used to judge particular actions ethically right or wrong; and within a particular political structure to justify publicly a course of conduct. Use (d) in this way is probably the readiest test of whether or not a particular rule is to be given normative status. For example, I can think of no part of our own society where a leader can say 'I did this because I enjoy ordering people about and I like to be famous': but he can say 'I did this for the common good.'
>
> Pragmatic rules are statements not about whether a particular line of conduct is just or unjust, but about whether or not it will be effective. They are normatively neutral. They may operate within the limits set by the rules of the game; or they may not. They range from rules of 'gamesmanship' (how to win without actually cheating) to rules which advise on how to win by cheating without being disqualified (what may be done, for example, on the 'blind' side of the referee in the boxing ring).

Bailey emphasizes how these two types of rules combine for effective action. Pragmatic rules are intended to highlight practical options and strategies available. In the case of university level technology planning, it requires an inquiry into all available needs and goals identified by university members regarding this technology planning initiative. Normative rules are needed to help provide a value gauge on what is ethically acceptable and unacceptable for university members regarding this technology planning initiative. Winning the game (the application of a successful technology planning initiative) depends on coordinated university efforts to understand the various stakeholder perspectives (students, faculty, administration, staff), along with an appreciation of how pragmatic and normative rules apply to the technology planning initiative.

Within this particular faculty level technology planning initiative, it is necessary to take a step back to see how each stakeholder group views the situation and what their values and objective are regarding the technology planning initiative. This may involve stakeholder surveys, strategic focus groups, public forums, and consultation. For this to be successful, a similar initiative must be replicated within the other university faculties and units that constitute the university as a whole. This will provide a thermometer reading of the actual temperature (values, interests, and needs) of the university as a social system. At the same time, it is important to have a balance between pragmatic and normative rule so that practical strategies do not take priority over university values and vice versa. This is a danger in large-scale university planning initiatives where there are limited resources and time. For instance, to address the high value placed on student interaction and learning, a course management type system may meet the normative values of all university stakeholders. However, the variation in the many course management type systems available requires additional university stakeholder consultation to get at the practical strengths and weakness of each system before committing to a solution. Using the practical utility of a course management type system as a band-aid to satisfy normative values of active learning engagement and cooperative learning could result in conflict (Which system is best and for whom?). That is why a balanced consideration of both normative and pragmatic aspects of the problem context must be explored within the faculty and between faculties. Based on these considerations, the following basic steps for guiding technoethical consultation are drawn from Luppicini (2010):

Step 1: Make sure the technological system is explicitly understood by all members within the design context. This includes system components and relations such as forms of technology mediation (and their impact).

Step 2: Perform technoethical assessment to discern all possible technical knowledge (facts and values) relevant to the design context.

Step 3: Focus design efforts on optimizing technological system operations to satisfy the needs and interests of all affected.

Step 4: Consult with representative from all groups affected by some technology in an effort to establish mutual understanding and agreement concerning important design issues.

In terms of the Vision 2015 planning strategy, it is recommended that other faculties within the university coordinate similar technology consultation initiatives to discuss at the university level. After this university level consultation is complete, each faculty will be more informed about the technology values, interests, and needs within the university as a whole. This could allow a more equitable technology solution for all university stakeholders to be applied.

One major limitation of this approach in the present context and other similar large-scale technology planning initiatives is that it requires time and effort to coordinate intra-university collaboration which is often difficult to achieve within university settings with demanding work schedules and deadlines faced by busy students, faculty members, and staff. However, it is the view of this author that anything worth doing is worth doing well and many of the excellent insights gleamed within this edited volume require an investment of time and effort to implement in a way that leverages university operations and satisfies its membership. In the end, it is necessary to recognize that technology is not the effortless solution to everything and all things. Instead, it is a complex system of tools and strategies that can be mobilized to advance teaching and learning within educational institutions. At the same time, it can also detract from the qualitative of teaching and learning within educational institutions. It is up to all of us to carefully research, plan, and implement technology in a way that adds value to education for all affected.

KEYWORDS

- **Committee meetings**
- **E-books**
- **Large-scale technology**
- **Misalignment**
- **Planning strategies**

References

1

Astrom, A., (2008). *E-learning quality aspects and criteria for evaluation of e-learning in higher education*. Stockholm: National Agency's Department of Evaluation.

Boot, E. W., van Merrienboer, J. J. G., & Theunissen, N. C. M. (2008). Improving the development of instructional software: Three building-block solutions to interrelate design and production. *Computers in Human Behavior, 24*, 1275–1292.

Buzzetto-More, N. A. (2008). Student perceptions of various e-learning components. *Interdisciplinary Journal of E-learning and Learning Objects, 4*, 113–135.

Campanella, S., Dimauro, G., Ferrante, A., Impedovo, D., Impedovo, S., Lucchese, M. G., ¼ Trullo, C. A. (2007). Engineering e-learning surveys: A new approach. *International Journal of Education and Information Technologies, 1*(2), 105–113.

Chorpothong, N., & Charmonman, S. (2004). An eLearning project for 100,000 students per year in Thailand. *International Journal of The Computer, the Internet and Management, 12*, 111–118.

Clark, D. (2004). *ADDIE—1975*. Retrieved from http://www.nwlink.com/%7Edonclark/history_isd/addie.html

Colston, R. (2008). *ADDIE model at learning-theories.com*. Retrieved from http://www.learning-theories.com/addie-model.html

Dick, W., & Carey, L. (1996). *The systematic design of instruction* (4th ed.). New York: Harper Collins.

Dilmaghani, M. (2003). *National providence and virtual education development capabilities in higher education*. Paper presented at the Virtual University Conference, Kashan, Iran.

E-learning. (2006). *Evaluation management*. Retrieved from http://www.elearning-engineering.com/evaluation.htm

Giveki, F. (2003). *Learning new methods in distance higher education*. Paper presented at the Virtual University Conference, Kashan, Iran.

Gladun, A., Rogushina, J., Garcıa-Sanchez, F., Martınez-Bejar, R., & Fernandez-Breis, J. T. (2009). An application of intelligent techniques and semantic web technologies in e-learning environments. *Expert Systems with Applications, 36*(2), 1922–1931.

Hejazi, A. (2007). *New wine-old bottles*. Retrieved from http://ictarticles.blogspot.com/2007/05/new-wine-old-bottles.html

Hung, T., Liu, C., Hung, C., Ku, H., & Lin, Y. (2007, September 3–7). *The establishment of an interactive e-learning system for engineering fluid flow and heat transfer*. Paper presented at the International Conference on Engineering Education—ICEE 2007, Coimbra, Portugal.

Jara, C. A., Candelas, F. A., Torres, F., Dormido, S., Esquembre, F., & Reinoso, O. (2009). Real-time collaboration of virtual laboratories through the Internet. *Computers & Education, 52*(1), 126–140.

Lee, J., & Lee, W. (2008). The relationship of e-learner's self-regulatory efficacy and perception of e-learning environmental quality. *Computers in Human Behavior, 24*, 32–47.

Littlejohn, A., Falconer, I., & Mcgill, L. (2008). Characterising effective eLearning resources. *Computers & Education, 50*, 757–771.

Magoha, P. W., & Andrew, W. O. (2004). The global perspectives of transitioning to e-learning in engineering education. *World Transactions on Engineering and Technology Education, 3*(2) 205–210.

Malachowski, M. J. (2002). *ADDIE based five-step method towards instructional design*. Retrieved from http://fog.ccsf.cc.ca.us/~mmalacho/OnLine/ADDIE.html

Morse, A., & Suktrisul, S. (2006). Introducing eLearning into secondary schools in Thailand [Special Issue]. *International Journal of The Computer, the Internet and Management, 14*, 81–85.

Motiwalla, L. F. (2007). Mobile learning: A framework and evaluation. *Computers & Education, 49*, 581–596.

Nelasco, S., Aruttharaj, A. N., & Alwinson, E. G. (2007). E-Learning for higher studies of India [Special Issue]. *International Journal of The Computer, the Internet and Management, 15*, 161–167.

Noori, M. (2003). *Traditional education or learning with computer.* Paper presented at the Virtual University Conference at Kashan Payame Noor College, Kashan, Iran.

Padilla-Melendez, A., Garrido-Moreno, A., & Aguila-Obra, A. R. (2008). Factors affecting e-collaboration technology use among management students. *Computers & Education, 51*, 609–623.

Pryor, C. R., & Bitter, G. G. (2008). Using multimedia to teach inservice teachers: Impacts on learning, application, and retention. *Computers in Human Behavior, 24*, 2668–2681.

Ramasundaram, V., Grunwald, S., Mangeot, A., Comerford, N. B., & Bliss, C. M. (2005). Development of an environmental virtual field laboratory. *Computers & Education, 45*, 21–34.

Ryder, M. (2008). *Instructional design models.* Retrieved from http://carbon.cudenver.edu/~mryder/ itc_data/idmodels.html#modern

Shee, D. Y., & Wang, Y. (2008). Multi-criteria evaluation of the web-based e-learning system: A methodology based on learner satisfaction and its applications. *Computers & Education, 50*, 894–905.

Siemens, G. (2002). *Instructional design in e-learning.* Retrieved from http://www.elearnspace.org/ Articles/InstructionalDesign.htm

Sirinaruemitr, P. (2004). Trends and forces for eLearning in Thailand. *International Journal of The Computer, the Internet and Management, 12*, 132–137.

Siritongthaworn, S., & Krairit, D. (2004). Use of interactions in e-learning: A study of undergraduate courses in Thailand. *International Journal of The Computer, the Internet and Management, 12*, 162–170.

Tabakov, S. (2008). E-learning development in medical physics and engineering. *Biomedical Imaging and Intervention Journal, 4*(1), e27.

UNESCO. (2006a). *UIS statics in brief-India.* Retrieved from http://stats.uis.unesco.org/unesco/TableViewer/document.aspx?ReportId=121&IF_Language=eng&BR_Country=3560

UNESCO. (2006b). *UIS statics in brief-Iran.* Retrieved from http://stats.uis.unesco.org/unesco/TableViewer/document.aspx?ReportId=121&IF_Language=eng&BR_Country=3640

Vate-U-Lan, P. (2007). Readiness of eLearning connectivity in Thailand [Special Issue]. *International Journal of The Computer, the Internet and Management, 15*, 21–27.

Willis, B. (2008). *Instance education in glance.* Retrieved from http://www.uiweb.uidaho.edu/eo/dist3.html

Witsawakiti, N., Suchato, A., & Punyabukkana, P. (2006). Thai language e-training for the hard of hearing [Special Issue]. *International Journal of The Computer, the Internet and Management, 14*, 41–46.

Zhiting, Z. (2004). The development and applications of eLearning technology standards in China. *International Journal of The Computer, the Internet and Management, 12*, 100–104.

2

Acar, A. (2008). Antecedents and consequences of online social networking behavior: The case of Facebook. *Journal of Website Promotion, 3*(1), 62–83.

Acquisti, A., & Gross, R. (2006). *Imagined communities: Awareness, information sharing, and privacy on the Facebook.* Paper presented at the Proceedings of Privacy Enhancing Technologies Workshop.

Boyd, D. (2006). Friends, Friendsters, and Top 8: Writing community into being on social network sites [Electronic Version]. *First Monday, 11.* Retrieved December 12, 2009, from http://firstmonday.org/htbin/cgiwrap/bin/ojs/index.php/fm/article/view/1418/1336

Boyd, D. (2007). *Why youth ♥ social network sites: The role of networked publics in teenage social life.* In D. Buckingham (Ed.), Youth, Identity, and Digital Media: M.I.T. Press.

Boyd, D. (2008). *Taken out of context: American teen sociality in networked publics.* Berkeley: University of California.

Boyd, D., & Ellison, N. B. (2008). Social network sites: Definition, history, and scholarship. *Journal of Computer-Mediated Communication, 13,* 210–230.

Buffardi, L. E., & Campbell, W. K. (2008). Narcissism and social networking web sites. *Personality and Social Psychology Bulletin, 34*(10), 1303–1314.

Christofides, E., Muise, A., & Desmarais, S. (2009). Information Disclosure and Control on Facebook: Are They Two Sides of the Same Coin or Two Different Processes? *CyberPsychology & Behavior, 12*(3), 341–345.

Coté, J. E. (2009). Identity formation and self development in adolescence. In R. M. Lerner & L. Steinberg (Eds.), *Handbook of adolescent psychology 1. Individual bases of adolescent development* (Vol. 1). Hoboken, NJ: Wiley.

Crook, C., & Harrison, C. (2008). *Web 2.0 Technologies for learning at key stages 3 and 4: Summary report.* London: BECTA.

Donath, J. (2008). Signals in social supernets. *Journal of Computer-Mediated Communication, 13*(1), 231–251.

Dong, Q., Urista, M. A., & Gundrum, D. (2008). The impact of emotional intelligence, self-esteem, and self-image on romantic communication over myspace. *CyberPsychology & Behavior, 11*(5), 577–578.

Ellison, N. B., Steinfield, C., & Lampe, C. (2007). The benefits of Facebook "friends:" social capital and college students' use of online social network sites [Electronic Version]. *Journal of Computer-Mediated Communication, 12*(4), article 1 from http://jcmc.indiana.edu/vol12/issue4/ellison.html

Fogel, J., & Nehmad, E. (2009). Internet social network communities: Risk taking, trust, and privacy concerns. *Computers in Human Behavior, 25*(1), 153–160.

Grasmuck, S., Martin, J., & Zhao, S. (2009). Ethno-racial identity displays on Facebook. *Journal of Computer-Mediated Communication, 15*(1), 158–188.

Greenhow, C., & Robelia, B. (2009a). Informal learning and identity formation in online social networks. *Learning, Media and Technology, 34*(2), 119–140.

Greenhow, C., & Robelia, B. (2009b). Old communication, new literacies: Social network sites as social learning resources. *Journal of Computer-Mediated Communication, 14*(4), 1130–1161.

Gross, R., Acquisti, A., & Heinz III, H. (2005). *Information revelation and privacy in online social networks.* Paper presented at the ACM workshop on Privacy in the electronic society.

Hargittai, E. (2007). Whose space? Differences among users and non-users of social network sites [Electronic Version]. *Journal of Computer-Mediated Communication, 13,* article 14 from http://jcmc.indiana.edu/vol13/issue1/hargittai.html

Hargittai, E. (2010). Digital na(t)ives? Variation in internet skills and uses among members of the "Net Generation". *Sociological Inquiry, 80*(1), 92–113.

Hargittai, E., Fullerton, L., Menchen-Travino, E., & Thomas, K. Y. (2010). Trust online: young adults' evaluation of web content. *International Journal of Communication, 4,* 468–494.

Hewitt, A., & Forte, A. (2006). *Crossing boundaries: Identity management and student/faculty relationships on the Facebook.* Paper presented at the CSCW, Banff, Alberta.

Hinduja, S., & Patchin, J. (2008). Personal information of adolescents on the Internet: A quantitative content analysis of MySpace. *Journal of Adolescence, 31,* 125–146.

Jones, S., Millermaier, S., Goya-Martinez, M., & Schuler, J. (2008). Whose space is MySpace? A content analysis of MySpace profiles. *First Monday, 13*(9).

Kleck, C., Reese, C., Behnken, D. Z., & Sundar, S. (2007). *The company you keep and the image you project.* Proceedings of the 57th annual conference for the ICA, 1–30.

Kuhn, D. (2009). Adolescent thinking. In R. M. Lerner & L. Steinberg (Eds.), *Handbook of adolescent psychology 1.* Individual bases of adolescent development (Vol. 1). Hoboken, NJ: Wiley.

Lampe, C., Ellison, N., & Steinfield, C. (2006). *A Face (book) in the crowd: Social searching vs. social browsing.* Proceedings of the 2006 20th anniversary conference on Computer supported cooperative work, 170.

Lampe, C., Ellison, N. B., & Steinfield, C. (2007). *A familiar face (book): Profile elements as signals in an online social network*. Proceedings of the SIGCHI conference on Human factors in computing systems, 444.

Lange, P. G. (2008). Publicly Private and Privately Public: Social Networking on YouTube. *Journal of Computer-Mediated Communication, 13*(1), 361–380.

Lenhart, A. (2007, December 18). *Teens and social media*. Retrieved November 8, 2009, from http://www.pewinternet.org/Reports/2007/Teens-and-Social-Media.aspx

Lenhart, A., & Madden, M. (2007a). *Social networking websites and teens*. Retrieved November 8, 2009, from http://www.pewinternet.org/Reports/2007/Social-Networking-Websites-and-Teens.aspx

Lenhart, A., & Madden, M. (2007b, September 11). *Teens, privacy & online social networks*. Retrieved November 8, 2009, from http://www.pewinternet.org/Reports/2007/Teens-Privacy-and-Online-Social-Networks.aspx

Lewis, J., & West, A. (2009). "Friending": London-based undergraduates' experience of Facebook. *New Media & Society, 11*(7), 1209–1229.

Lewis, K., Kaufman, J., & Christakis, N. (2008). The taste for privacy: An analysis of college student privacy settings in an online social network. *Journal of Computer-Mediated Communication, 14*(1), 79–100.

Livingstone, S. M. (2008). Taking risky opportunities in youthful content creation: teenagers' use of social networking sites for intimacy, privacy and self-expression. *New Media & Society, 10*(3), 393–411.

Madge, C., Meek, J., Wellens, J., & Hooley, T. (2009). Facebook, social integration and informal learning at university: "It is more for socialising and talking to friends about work than for actually doing work." *Learning, Media and Technology, 34*(2), 141–155.

Mallan, K., & Giardina, N. (2009). Wikidentities: Young people collaborating on virtual identities in social network sites. *First Monday, 14*(6).

Manago, A. M., Graham, M. B., Greenfield, P. M., & Salimkhan, G. (2008). Self-presentation and gender on MySpace. *Journal of Applied Developmental Psychology, 29*(6), 446–458.

Marwick, A. E. (2008). To catch a predator? The MySpace moral panic. *First Monday,* 13(6).

Mazer, J. P., Murphy, R. E., & Simonds, C. J. (2007). I'll see you on "Facebook": The effects of computer-mediated teacher self-disclosure on student motivation, affective learning, and classroom climate. *Communication Education, 56*(1), 1–17.

Mazer, J. P., Murphy, R. E., & Simonds, C. J. (2008). The Effects of Teacher Self-Disclosure via "Facebook" on Teacher Credibility. RCA Vestnik (Russian Communication Association), 30–37.

Mueller, J., Wood, E., & Wiloughby, T. (2008). The integration of computer technology in the classroom. In T. Willoughby & E. Wood (Eds.), *Children's learning in a digital world*. Malden, MA: Blackwell.

Muise, A., Christofides, E., & Desmarais, S. (2009). More information than you ever wanted: Does Facebook bring out the green-eyed monster of jealousy? *CyberPsychology & Behavior, 12*(4), 441–444.

Ofcom. (2008). *Ofcom's Submission to the Byron Review (Annex 5: The Evidence Base—The Views of Children, Young People and Parents)*. Retrieved November 8, 2009, from http://www.ofcom.org.uk/research/telecoms/reports/byron/annex5.pdf

Orr, E. S., Sisic, M., Ross, C., Simmering, M. G., Arseneault, J. M., & Orr, R. R. (2009). The influence of shyness on the use of Facebook in an undergraduate sample. *CyberPsychology & Behavior, 12*(3), 337–340.

Pasek, J., More, E., & Hargittai, E. (2009). Facebook and academic performance: Reconciling a media sensation with data. *First Monday, 14*(5).

Peluchette, J., & Karl, K. (2008). Social networking profiles: an examination of student attitudes regarding use and appropriateness of content. *CyberPsychology & Behavior, 11*(1), 95–97.

Pempek, T. A., Yermolayeva, Y. A., & Calvert, S. L. (2009). College students' social networking experiences on Facebook. *Journal of Applied Developmental Psychology, 30*(3), 227–238.

Pfeil, U., Arjan, R., & Zaphiris, P. (2009). Age differences in online social networking: A study of user profiles and the social capital divide among teenagers and older users in MySpace. *Computers in Human Behavior, 25*(3), 643–654.

Raacke, J., & Bonds-Raacke, J. (2008). MySpace and Facebook: Applying the uses and gratifications theory to exploring Friend-networking sites. *CyberPsychology & Behavior, 11*(2), 169–174.

Rosen, L. D. (2006, August 14). *Adolescents in MySpace—Executive Summary*. Retrieved November 8, 2009, from http://www.csudh.edu/psych/Adolescents%20in%20MySpace%20-%20Executive%20Summary.pdf

Rosen, L. D., Cheever, N. A., & Carrier, L. M. (2008). The association of parenting style and child age with parental limit setting and adolescent MySpace behavior. *Journal of Applied Developmental Psychology, 29*(6), 459–471.

Ross, C., Orr, E. S., Sisic, M., Arseneault, J. M., Simmering, M. G., & Orr, R. R. (2009). Personality and motivations associated with Facebook use. *Computers in Human Behavior, 25*(2), 578–586.

Schrock, A. (2008). Examining social media usage: Technology clusters and social network site membership. *First Monday, 14*(1).

Seder, J. P., & Oishi, S. (2009). Ethnic/racial homogeneity in college students' Facebook friendship networks and subjective well-being. *Journal of Research in Personality, 43*(3), 438–443.

Selwyn, N. (2009). Faceworking: Exploring students' education-related use of Facebook. *Learning, Media and Technology, 34*(2), 157–174.

Sharples, M., Graber, R., Harrison, C., & Logan, K. (2009). E-safety and Web 2.0 for children aged 11–16. *Journal of Computer Assisted Learning, 25*(1), 70–84.

Sheldon, P. (2008a). Student Favorite: Facebook and Motives for its use. *Southwestern Mass Communication Journal, 23*(2), 39–53.

Sheldon, P. (2008b). The relationship between unwillingness-to-communicate and students' Facebook use. *Journal of Media Psychology: Theories, Methods, and Applications, 20*(2), 67–75.

Silverstone, R. (2007). Media and morality: On the rise of the mediapolis. Cambridge/Malden, MA: Polity Press.

Smetana, J. G., & Villalobos, M. (2009). Social cognitive development in adolescence. In R. M. Lerner & L. Steinberg (Eds.), *Handbook of adolescent psychology 1. Individual bases of adolescent development (Vol. 1)*. Hoboken, NJ: Wiley.

Steinfield, C., Ellison, N. B., & Lampe, C. (2008). Social capital, self-esteem, and use of online social network sites: A longitudinal analysis. *Journal of Applied Developmental Psychology, 29*(6), 434–445.

Strano, M. (2008). User descriptions and interpretations of self-presentation through Facebook profile images. *Cyberpsychology: Journal of Psychosocial Research on Cyberspace, 2*(2).

Subrahmanyam, K., Reich, S. M., Waechter, N., & Espinoza, G. (2008). Online and offline social networks: Use of social networking sites by emerging adults. *Journal of Applied Developmental Psychology, 29*(6), 420–433.

Takahashi, T. (2008). Mobile phones and social networking sites: Digital natives' engagement with media in everyday life in Japan. *Media, Communication and Humanity,* 21–23.

Thelwall, M. (2008a). Fk yea I swear: cursing and gender in MySpace. *Corpora, 3*(1), 83–107.

Thelwall, M. (2008b). Social networks, gender, and friending: An analysis of MySpace member profiles. *Journal of the American Society for Information Science and Technology, 59*(8), 1321–1330.

Tong, S., Van Der Heide, B., Langwell, L., & Walther, J. (2008). Too much of a good thing? The relationship between number of friends and interpersonal impressions on Facebook. *Journal of Computer-Mediated Communication, 13*(3), 531–549.

Tufekci, Z. (2008). Grooming, gossip, Facebook and Myspace: What can we learn about these sites from those who won't assimilate? *Information, Communication & Society, 11*(4), 544–564.

Valenzuela, S., Park, N., & Kee, K. F. (2009). Is there social capital in a social network site?: Facebook use and college students' life satisfaction, trust, and participation. *Journal of Computer-Mediated Communication, 14*(4), 875–901.

Valkenburg, P., Peter, J., & Schouten, A. (2006). Friend networking sites and their relationship to adolescents' well-being and social self-esteem. *CyberPsychology & Behavior, 9*(5), 584–590.

Walther, J., Van Der Heide, B., Hamel, L., & Shulman, H. (2009). Self-generated versus other-generated statements and impressions in computer-mediated communication: A test of warranting theory using Facebook. *Communication Research, 36*(2), 229–253.

Walther, J., Van Der Heide, B., Kim, S., Westerman, D., & Tong, S. (2008). The role of friends' appearance and behavior on evaluations of individuals on Facebook: Are we known by the company we keep? *Human Communication Research, 34*(1), 28–49.

Wang, S. S., Moon, S., Kwon, K. H., Evans, C. A., & Stefanone, M. A. (2009). Face off: Implications of visual cues on initiating friendship on Facebook. *Computers in Human Behavior,* 226–234.

Wartella, E., & Jennings, N. (2000). Children and computers: New technology—old concerns. *The future of Children: Children and Computer Technology,* 10(2)

Wartella, E., & Reeves, B. (1985). Historical trends in research on children and the media: 1900–1960. *Journal of Communication, 35*(2), 118–133.

West, A., Lewis, J., & Currie, P. (2009). Students' Facebook "friends": public and private spheres. *Journal of Youth Studies, 12*(6), 615–627.

Williams, A., & Merten, M. (2008). A review of online social networking profiles by adolescents: Implications for future research and intervention. *Adolescence, 43*(170), 253–274.

Wing, C. (2005). Phase II student survey. Young Canadians in a wired world. Retrieved March 5, 2008, from www.media-awareness.ca

Wolak, J., Finkelhor, D., Mitchell, K. J., & Ybarra, M. L. (2008). Online "predators" and their victims: Myths, realities, and implications for prevention and treatment. *American Psychologist, 63*(2), 111–128.

Ybarra, M., & Mitchell, K. (2008). How risky are social networking sites? A comparison of places online where youth sexual solicitation and harassment occurs. *Pediatrics, 121*(2), e350.

Zhao, S. (2009). Teen adoption of MySpace and IM: Inner-city versus suburban differences. *CyberPsychology & Behavior, 12*(1), 55–58.

Zhao, S., Grasmuck, S., & Martin, J. (2008). Identity construction on Facebook: Digital empowerment in anchored relationships. *Computers in Human Behavior, 24*(5), 1816–1836.

Zywica, J., & Danowski, J. (2008). The faces of Facebookers: Investigating social enhancement and social compensation hypotheses; predicting Facebook™ and offline popularity from sociability and self-esteem, and mapping the meanings of popularity with semantic networks. *Journal of Computer-Mediated Communication, 14*(1), 1–34.

3

Alexander, C. (2008). An overview of LAMS (Learning Activity Management System). *Teaching English with Technology Journal, 8*(3), 1–22.

Beatty, K. (2003). *Teaching and researching computer-assisted language learning.* Harlow: Pearson Education.

Dalziel, J. (2003). *Implementing learning design: The Learning Activity Management System (LAMS).* Proceedings of the 20th Annual Conference of the Australian Society for Computers in Learning in Tertiary Education (ASCILITE).

Fan, Z., & Jiaheng, C. (2008). Learning activity sequencing in personalized education system. *Journal of Natural Science, 13*(4).

Mann, S. (2008). The problems of online collaboration for junior high school students: Can the Learning Activity Management System (LAMS) benefit students to learn via online learning? Australia: CoCo Research Centre, University of Sydney.

Nor Azan, M. Z. (2007). Learning Activities Management System (SPAP). *Proceedings of International Conference on Electrical Engineering and Informatics.* Institut Teknologi Bandung, Indonesia.

FURTHER READING

Alexander, C. (2007a). Using the Internet in TESOL14. In S. Walker, M. Ryan, & R. Teed (Eds.), *Proceeding of Design for Learning. 2007* (pp 28–34). Greenwich University. Retrieved from http://web-dev-csc.gre.ac.uk/conference/conf32/index.php?p=246

Alexander, C. (2007b). A case study of English language teaching using the Internet in Intercollege's language laboratory. *The International*

Journal of Technology, Knowledge and Society, 3(1), 1–15. Retrieved from http://christopherandrewalexander.cgpublisher.com/

Britain, S. (2004). *A Review ofILearning design: Concept, specifications and tools: A report for the JISC e-learning pedagogy programme.* Retrieved May 20, 2008, from http://www.jisc.ac.uk/uploaded_documents/ACF1ABB.doc.

Cohen, R. F., Meacham, A., & Skaff, J. (2006). *Teaching graphs to visually impaired students using an active auditory interface.* Proceedings of the 37th SIGCSE Technical Symposium on Computer Science Education (pp. 279–282). Houston, TX.

Crescenzi, P., & Nocentini, C. (2007). *Fully integrating algorithm visualization into a CS2 course: A two-year experience.* Proceedings of the 12th Annual Conference on Innovation and Technology in Computer Science Education (ITiCSE'07) (pp. 296–300).

Curzon, P., & McOwan, P. W. (2008). Engaging with computer science through magic shows. *ACM SIGCSE Bulletin, 40*(3), 179–183.

Dalziel, J. (2003a). Discussion paper for learning activities and meta-data. *Macquarie E-learning Centre of Excellence (MELCOE).* Retrieved May 10, 2008, from http://www.lamsinternational.com/documents/LearningActivities.Metadata.Dalziel.pdf.

Erkan, A., VanSlyke, T. J., & Scaffidi, T. (2007). *Data Structure Visualization with LaTeX and Prefuse.* Dundee: ITICSE.

Lauer, T. (2006). Learner interaction with algorithm visualizations: Viewing vs. changing vs. constructing. *ACM SIGCSE Bulletin, 38*(3).

Levy, P., Aiyegbayo, O., Little, S., Loasby, I., & Powell, A. (2008). *Designing and sharing inquiry-based learning activities: LAMS evaluation case study.* (Centre for Inquiry-based Learning in the Arts and Social Sciences, University of Sheffield). Retrieved June 20, 2008, from http://www.jisc.ac.uk/media/documents/programmes/elearningpedagogy/desilafinalreport.pdf

Myller, N., Laakso, M., & Korhonen, A. (2007). Analyzing engagement taxonomy in collaborative algorithm visualization. *ITiCSE,* 251–255.

Russell, T., Varga-Atkins, T., & Roberts, D. (2005). Learning activity management system specialist schools trust pilot. *A Review for Becta*

and the Specialist Schools and Academies Trust by CRIPSAT. Centre for Lifelong Learning: University of Liverpool. Retrieved March 10, 2008, from http://www.cripsat.org.uk/current/elearn/bectalam.htm

Schweitzer, D., & Brown, W. (2007). Interactive visualization for the active learning classroom. *ACM SIGCSE Bulletin, 39*(1).

Subramanian, K. R., & Cassen, T. (2008). A cross-domain visual learning engine for interactive generation of instructional materials. *ACM SIGCSE Bulletin, 40*(1).

4

Al-Jumaily, A., & Stonyer, H. (2000). Beyond teaching and research: Changing engineering academic work. *Global Journal of Engineering Education, 4*(1), 89–97.

Baron, R. (1998). What type am I?: Discover who you really are. New York: Penguin Putnam.

Boyer, E. L. (1990). *Scholarship reconsidered: Priorities of the professoriate.* Stanford, CA: Carnegie Foundation for the Advancement of Teaching.

Coleman, D. J. (1998). *Applied and academic geomatics into the 21st Century.* Paper presented at the FIG Commission 2, The XXI Inter. FIG Congress, Brighton, England.

CSU. (2004). *Chico distance & online education.* Retrieved from http://rce.csuchico.edu/online/site.asp

Dale, E. (1969). *Audiovisual methods in teaching* (3rd ed.). New York: Dryden Press.

Dunn, R. (1990). Understanding the Dunn and Dunn learning styles model and the need for individual diagnosis and prescription. *Reading, Writing and Learning Disabilities, 6,* 223–247.

Dunn, R. (1992). Learning styles network mission and belief statements adopted. *Learning Styles Network Newsletter, 13*(2), 1.

Dunn, R., Beaudry, J. S., & Klavas, A. (1989). Survey of research on learning styles. *Educational Leadership, 46*(6), 50–58.

Felder, R. M. (1993). Reaching the second tier: Learning and teaching styles in college science education. *Journal of College Science Teaching, 23*(5), 286–290.

Finelli, C. J., Klinger, A., & Budny, D. D. (2001). Strategies for improving the classroom environment. *Journal of Engineering Education, 90*(4), 491–501.

Foertsch, J., Moses, G., Strikwerda, J., & Litzkow, M. (2002). Reversing the lecture/homework paradigm using e-TEACH Web-based streaming video software. *Journal of Engineering Education, 91*(3), 267–275.

Forks, G. (2009). *UND online & distance education.* Retrieved from http://distance.und.edu/

Frances, C., Pumerantz, R., & Caplan, J. (1999). Planning for instructional technology: What you thought you knew could lead you astray. *Change, 31*(4), 24–33.

Hinchcliff, J. (2000). Overcoming the anachronistic divide: Integrating the why into the what in engineering education. *Global Journal of Engineering Education, 4*(1), 13–18.

Johnson, D. W., Johnson, R. T., & Smith, K. A. (1991). *Active learning: Cooperation in the college classroom.* Edina, MN: Interaction Book Company.

Johnson, D. W., Johnson ,R. T., & Smith, K. A. (1992). Cooperative learning: Increasing college faculty instructional productivity. *ERIC Digest.* Washington, DC: The George Washington University, School of Education and Human Development.

Jung, C. G. (1971). *Psychological types* .Princeton, NJ: University of Arkansas Press.

Kolb, D. A. (1981). *Learning styles and disciplinary differences.* San Francisco, CA: Jossey-Bass.

Kolb, D. A. (1984). *Experiential learning: Experience as the source of learning and development* (1st ed.). Englewood Cliffs, NJ: Prentice-Hall.

Kolmos, A. (1996). Reflections on project work and problem-based learning. *European Journal of Engineering Education, 21*(2), 141–148.

Mamchur, C. (1990). Cognitive type theory and learning style, Association for Supervision and Curriculum Development. *A teacher's guide to cognitive type theory and learning style.* Alexandria, VA: Association for Supervision and Curriculum Development.

Myers, L. B., & McCaulley, M. H. (1985). *Manual-guide to the development and use of the Myers-Briggs indicator.* Palo Alto, CA: Consulting Psychologists Press.

Oprean, C. (2006). The Romanian contribution to the development of engineering education. *Global Journal of Engineering Education, 10,* 45–50.

Overview of e3L. (2005). Retrieved from http://e3learning.edc.polyu.edu.hk/main.htm

Purcell-Roberston, R. M., & Purcell, D. F. (2000). Interactive distance learning. *Distance learning technologies: Issues, trends and opportunities.* Hershey, PA: IDEA Publishing Group.

Rugarcia, A., Felder, R. M., Woods, D. R., & Stice, J. E. (2000). The future of engineering education I: A vision for a new century. *Chemical Engineering Education, 349*(1), 16–25.

UI. (2004). *Engineering outreach distance education.* Retrieved from http://www.uidaho.edu/evo

Wild, R. H., Griggs, K. A., & Downing, T. (2002). A framework for e-learning as a tool for knowledge management. *Industrial Management and Data Systems, 102*(7), 371–381.

5

ACS. (2009). *ACS complete short course listings.* Retrieved from http://portal.acs.org/portal/acs/corg/content?_nfpb=true&_pageLabel=PP_ARTICLEMAIN&node_id=273&content_id=CTP_006408&use_sec=true&sec_url_var=region1&__uuid=f3138678-d8ec-4b90-83a7-781b4b80b9bf

Allen, M. W. (2003). *Michael Allen's guide to e-learning: Building interactive, fun, and effective learning programs for any company.* Hoboken, NJ: John Wiley & Sons.

Baker, A., Navarro, E. O, & van der Hoek, A. (2005). An experimental card game for teaching software engineering processes. *Journal of System and Software, 75,* 3–16.

Bergstrom, S. (2001). *CMI guidelines for interoperability AICC.* Retrieved from http://www.aicc.org/docs/tech/cmi001v3-5.pdf

Chatterjee, A. (2005). Mathematics in engineering. *Current Science, 88*(3), 405–414.

Conn, R. (2002). Developing software engineers at the C-130J software factory. *IEEE Software, 19*(5), 25–29.

Connolly, T., & Stansfield, M. (2007). *Principles of effective online teaching*. Santa Rosa, CA: Informing Science Press.

DCMI. (2009). Dublin Core Metadata Element Set (Version 1.1). Retrieved from http://dublincore.org/documents/dces/

Digitalthink. (2003). *SCORM™: The e-learning standard*. Retrieved from http://www.adlnet.org/index.cfm?fuseaction=scormabt

Flowers, M., & Ching, H. W. (2008). *Lecture on carnot cycle*. Retrieved from http://galileoandeinstein.physics.virginia.edu/more_stuff/flashlets/carnot.htm

Friesen, N. (2004). *The e-learning standardization landscape*. Retrieved from http://www.cancore.ca/docs/intro_e-learning_standardization.html

Gottschalk, T. H. (2009). *Distance education at a glance*. Retrieved from http://www.uiweb.uidaho.edu/eo/distglan

Grunbacher, P., Seyff, N., Briggs, R. D., In, H. P., & Kitapci, H., & Port, D. (2007). Making every student a winner: The win-win approach in software engineering education. *Journal of System and Software, 80*, 1191–1200.

Harper, K. C., Chen, K., & Yen, D. C. (2004). Distance learning, virtual classrooms, and teaching pedagogy in the Internet environment. *Technology in Society, 26*, 585–598.

Hirsch, B. E., Thoben, K. D., & Hoheisel, J. (1998). Requirements upon human competencies in globally distributed manufacturing. *Computers in Industry, 36*, 49–54.

Horton, W. (2001). *Designing web-based training: How to teach anyone anything anywhere anytime*. New York: John Wily & Sons.

Hwang, F. (2004*). NTNUJAVA virtual physics laboratory java simulations in physics: Enjoy the fun!* Retrieved from http://www.phy.ntnu.edu.tw/ntnujava/index.php?PHPSESSID=61e16b595f51a79534ea86dd507e6fce&topic=23.msg156#msg156

IEEE. (2005). 1484.12.1: *IEEE standard for learning object metadata*. Retrieved from http://ltsc.ieee.org/wg12/

Imperial-College-London. (2009). *Information & communication technologies (ICT): An e-learning glossary*. Retrieved from http://www3.imperial.ac.uk/ict/services/teachingandresearchservices/elearning/aboutelearning/elearningglossary

IMS. (2009). *Innovation adoption learning*. Retrieved from http://www.imsglobal.org/specifications.html

Ion, D. (2007). *Physics and engineering*. Retrieved from http://www.cs.sbcc.cc.ca.us/~physics/

Kaufman, D. B., Felder, R. M., & Fuller, H. (2000). Accounting for individual effort in cooperative learning teams. *Journal of Engineering Education, 89*(2), 133–140.

Kilby, T. (2009). *What is web-based training?* Retrieved from http://www.webbasedtraining.com/primer_whatiswbt.aspx

Le@rningFederation (2008). *Metadata elements comparison: Vetadata and ANZ-LOM*. Retrieved from http://www.thelearningfederation.edu.au/verve/_resources/ANZLOM-VETADATA-comparison-v1-0.pdf

Mathias, L., Frantz, P., Brust, G., Montogery, J., Smith, V., Simmons, M., Wilson, R., Orna, M. V., Whitfield, J. M., Droske, J., & Badger, R. (2009). *Polymer science learning center*. from http://pslc.ws/macrog.htm

Monahan, T., McArdle, G., & Bertolotto, M. (2008). Virtual reality for collaborative e-learning. *Computers & Education, 50*, 1339–1353.

Pyrz, R., Olhoff, N., Lund, E., & Thomsen, O. T. (2003). *Mechanic for*. Retrieved from http://me.aau.dk/GetAsset.action?contentId=1987778&assetId=3356989

Ryder, M. (2008). *Instructional design models*. Retrieved from http://carbon.cudenver.edu/~mryder/itc_data/idmodels.html#modern

Smith, R. J. (2009). *Engineering*. Retrieved from http://www.britannica.com/EBchecked/topic/187549/engineering

Wintermantel., K. (1999). Process and product engineering achievements, present and future challenges. *Chemical Engineering Science, 54*, 1601–1620.

YilDirim, S. (2009). *Web-based training: Design and implementation issues*. Retrieved from http://ocw.metu.edu.tr/informatics-institute/web-based-training-design-and-implementation

6

Alvesson, M., & Sköldberg, K. (2009). *Reflexive methodology: New vistas for qualitative research* (2nd ed.). Los Angeles, CA: Sage.

Ayman-Nolley, S. (1992). Vygotsky's perspective on the development of imagination and creativity. *Creativity Research Journal, 5*, 77–85.

Barone, T. (2001). Science, art, and the predispositions of educational researchers. *Educational Researcher, 30*, 24–28.

Bullough, R. V., & Pinnegar, S. (2001). Guidelines for quality in autobiographical forms of self-study research. *Educational Researcher, 30*, 13–21.

Butler-Kisber, L., & Borgerson, J. (1997). *Alternative representation in qualitative inquiry: A student-instructor retrospective*. Paper presented at the Annual Meeting of the American Educational Research Association, Chicago, IL.

Creswell, J. W. (2007). *Qualitative inquiry & research design: Choosing among five approaches* (2nd ed.). Thousand Oaks, CA: Sage.

De Bono E. (1970). *Lateral thinking: Creativity step by step*. New York: Harper and Row.

Finley, S., & Knowles, J. G. (1995). Researcher as artist/Artist as researcher. *Qualitative Inquiry, 1*, 110–142.

Hamilton, M. L., & Pinnegar, S. (2009). Creating representation: Using collage in self-study. In D. L. Tidwell, M. L. Heston & L. M. Fitzgerald (Eds.), *Research methods for the self-study of practice* (pp. 155–170). Dordrecht: Springer.

Leavy, P. (2009). *Method meets art: Arts-based research practice*. New York: Guilford Press.

Leitch, R. (2006). Limitations of language: Developing arts-based creative narrative in stories of teachers' identities. *Teachers and Teaching: Theory & Practice, 12*, 549–569.

Lindqvist, G. (2003). Vygotsky's theory of creativity. *Creativity Research Journal, 15*, 245–251.

McNiff, S. (2007). Art based research. In *Knowles Handbook* (pp. 29–40). Retrieved from

http://www.corwin.com/upm-data/18069_Chapter_3.pdf

Mullen, C. A. (2000). Linking research and teaching: A study of graduate student engagement. *Teaching in Higher Education, 5*, 5–21.

Samaras, A. P. (2009). Explorations in using arts-based self-study methods. *International Journal of Qualitative Studies in Education, 23*(6), 719–736.

Schön, D. (1983) *The reflective practitioner: How professionals think in action*. London: Temple Smith.

Smolucha, F. (1992). A reconstruction of Vygotsky's theory of creativity. *Creativity Research Journal, 5*, 49–67.

Sullivan, G. (2006). Research acts in art practice. *Studies in Art Education, 48*, 19–35.

FURTHER READING

Barone, T. (1995). The purposes of arts-based educational research. *International Journal of Educational Research, 23*, 169–180.

Barone, T., & Eisner, E. (2006). Arts-based educational research. In J. L. Green, G. Camilli & P. B. Elmore (Eds.), *Handbook of complementary methods in education research* (pp. 95–110). Mahwah, NJ: Lawrence Erlbaum Associates.

Chenail, R. J. (2008). "But is it research?": A review of Patricia Leavy's method meets art: Arts-based research practice. *The Weekly Qualitative Report, 1*(2), 7–12.

Eisner, E. W. (2004). *Artistry in teaching: Cultural Commons*. Retrieved February 1, 2009, from http://www.culturalcommons.org/eisner.htm

Ham, V., & Kane, R. (2007). Finding a way through the swamp: A case for self-study as research. In J. J. Loughran, M. L. Hamilton, V. Kubler LaBoskey, & T. Russell (Eds.), *International handbook of self-study of teaching and teacher education practices* (pp. 103–150). Dordrecht: Springer.

Piantanida, M., McMahon, P. L., & Garman, N. B. (2003). Sculpting the contours of arts-based educational research within a discourse community. *Qualitative Inquiry, 9*, 182–191.

Sava, I., & Nuutinen, K. (2003). At the meeting place of word and picture: Between art and inquiry. *Qualitative Inquiry, 9*, 515–534.

Simons, H., & Hicks, J. (2006). Opening doors: Using the creative arts in learning and teaching. *Arts & Humanities in Higher Education, 5*, 77–90.

Suominen, A. (2006). Writing with photographs writing self: Using artistic methods in the investigation of identity. *International Journal of Education Through Art, 2*, 139–156.

Vaughan, K. (2005). Pieced together: Collage as an artist's method for interdisciplinary research. *International Journal of Qualitative Methods, 4*, 1–21.

7

Anohiina, A. (2005). Analysis of the terminology used in the field of virtual learning. *Educational Technology & Society, 8*, 91–102.

Bouras, C., Philopoulos, A., & Tsiatsos, T. (2001). E-learning through distributed virtual environments. *Journal of Network and Computer Applications, 24*, 175–199.

Chen, R., & Hsiang, C. (2007). A study on the critical success factors for corporations embarking on knowledge community-based e-learning. *Information Sciences, 177*, 570–586.

Chittaro, L., & Ranon, R. (2007). Web3D technologies in learning, education and training: Motivations, issues, opportunities. *Computers & Education, 49*, 3–18.

Cockburn, A., & Mckenzie, B. (2004). Evaluating spatial memory in two and three dimensions. *International Journal of Human-Computer Studies, 61*, 359–373.

Dalgarno, B., & Hedberg, J. (2001). *3D learning environments in tertiary education.* Paper presented at the 18th annual conference of the Australasian Society for Computers in Learning in Tertiary Education, Melbourne, Australia.

De Lucia, A., Francese, R., Passero, I., & Tortora, G. (2008). Development and evaluation of a virtual campus on Second Life: The case of SecondDMI. *Computers & Education*, in press.

Gervasi, O., Riganelli, A., Pacifici, L., & Laganà, A. (2004). VMSLab-G: A virtual laboratory prototype for molecular science on the Grid. *Future Generation Computer Systems, 20*, 717–726.

Hamza-Lup, F. G., & Stefan, V. (2007). *Web 3D & Virtual Reality: Based Applications for Simulation and e-learning.* Paper presented at the 2nd International Conference on Virtual Learning, ICVL, Constanta, Romania.

Harper, K. C., Chen, K., & Yen, D. C. (2004). Distance learning, virtual classrooms and teaching pedagogy in the Internet environment. *Technology in Society, 26*, 585–598.

Hung, T. C., Wang, S. K., Tai, S. W., & Hung, C. T. (2007). An innovative improvement of engineering learning system using computational fluid dynamics concept. *Computers & Education, 48*, 44–58.

Jonasson, J. (2005). *3D learning environment. Will it add the 3rd dimension to e-learning and teaching?* Paper presented at the 3rd International Conference on Multimedia and ICTs in Education, Cáceres, Spain.

Lee, J., Hong, N. L., & Ling, N. L. (2002). An analysis of students' preparation for the virtual learning environment. *Internet and Higher Education, 4*, 231–242.

Ling, C., Gen-Cai, C., Chen-Guang, Y., & Chuen, C. (2003). International Conference on Communication Technology proceeding, 2, 1655–1661.

Liu, C. C., Hung, C. T., Hung, T. C., Pei, B. S., & Zhang, L. (2006). *The development of an innovative interactive e-learning system in computational thermal-hydraulics for engineers.* Paper presented at the EDMEDIA, Chesapeake, VA, AACE.

Mangina, E., & Kilbride, J. (2008). Utilizing vector space models for user modeling within e-learning environments. *Computers & Education, 51*, 493–505.

Martins, L. L., & Kellermanns, F. W. (2004). A model of business school students' acceptance of a web-based course management system. *Academy of Management Learning and Education, 3*, 7–26.

Monahan, T., McArdle, G., & Bertolotto, M. (2008). Virtual reality for collaborative e-learning. *Computers & Education, 50*, 1339–1353.

Olivier, B. A., & Liver, O. (2003). Learning content interoperability standards. In A. Littlejohn

(Ed.), *Reusing online resources*. London: Kogan Page.

Ramasundaram, V., Grunwald, S., Mangeot, A., Comerford, N. B., & Bliss, C. M. (2005). Development of an environmental virtual field laboratory. *Computers & Education, 45*, 21–34.

Sperka, M. (2004). *Web 3D and new forms of human computer interaction*. Paper presented at the 2nd International Symposium of Interactive Media Design, Istanbul, Turkey.

van Raaij, E. M., & Schepers, J. J. L. (2008). The acceptance and use of a virtual learning environment in China. *Computers & Education, 50*, 838–852.

Zio, E. (2009). Reliability engineering: Old problems and new challenges. *Reliability Engineering and System Safety, 94*, 125–141.

Morris C. W. (1956).—*Varieties of human value*. Chicago, IL: University of Chicago Press.

Oishi, S., Schimmack, U., Diener, E. & Suh, E. (1998). The measurement of values and individualism collectivism. *Personality and Social Psychology Bulletin, 24*(11), 1177–1189.

Parsons, T. & Shils, E. (Eds.) (1951). *Towards a general theory of action*. Cambridge, MA: Harvard University Press.

Rohan, M. J. (2000). A rose by any name? The values construct. *Personality and Social Psychology Review, 4*,255–277.

Verplanken, B. & Holland, R. W. (2002). Motivated decision making: Effects of activation and self-centrality of values on choices and behavior. *Journal of Personality and Social Psychology, 82*(3), 434–447.

8

Allport, G. W. (1961). *Pattern and growth in personality*. New York: Holt, Rinehart & Winston.

Bardi, A. (2000). *Relations of values to behavior in everyday situations*. Unpublished doctoral dissertation. Jerusalem: The Hebrew University.

Bardi, A., & Schwartz S. H. (2003). Values and behavior: Strength and structure of relations. *Personality and Social Psychology Bulletin, 29*(10), 1207–1220.

Bilsky, W., & Schwartz, S. H. (1987). Toward a universal psychological structure of human values. *Journal of Personality and Social Psychology, 53*(3), 550–562.

Boudon, R. (2001). *The origins of values: Essays in the sociology and philosophy of beliefs*. New York: Transaction Publishers.

Davidov, E., Schmidt, P., & Schwartz, S. H. (2008). Bringing values back in. *Public Opinion Quarterly, 72*(3), 420–445.

Eisner, E. (1997). The promise and perils of alternative forms of data representation. *Educational Researcher, 26*(6), 4–10.

Kohn, M. L. (1977). *Class and conformity: A study in values*. Chicago, IL: The University of Chicago Press.

Maslow A. H. (1965).—*Eupsychian management*. Homewood, IL: Dorsey Press.

9

Academhack.com (2008). *Twitter for academia*. Retrieved August 24, 2010, from http://academhack.outsidethetext. com/home/2008/twitter-for-academia/

Butler-Kisber, L. (2007). Collage as inquiry. *Knowles handbook* (pp. 265–276, Chap. 22).

Currie, A. W. (2009). Students' Facebook 'friends': Public and private spheres. *Social Policy,* 615–627.

Eckartsberg, V. (1986). *Life-world experience: Existential-phenomenological research approaches in psychology*. Lanham, MD: University Press of America.

Ellison, N., Steinfield, C., & Lampe, C. (2007). The benefits of Facebook "friends": Social capital and college students' use of online social network sites. *Journal of Computer-Mediated Communication, 12*(4).

Gershon, L. (2008). Email my heart: Remediation and romantic break-ups. *Anthropology Today, 24*(6), 13–15. Retrieved from Academic Search Complete database.

Greenfield, P., & Subrahmanyam, K. (2008). Online communication and adolescent relationships. [Electronic version]. *The Future of Children, 18*, 119–146.

Massimini, M., & Peterson, M. (2009). Information and communication technology: Affects on

U.S. college students. *Cyberpsychology, 3*(1), 1–12.

Matsuba, M. (2006). Searching for self and relationships online. *CyberPsychology & Behavior, 9*(3), 275–284. Retrieved from Academic Search Complete database.

McNiff, S. (1998). *Art-based research*. Philadelphia, PA: Athenauem Press.

Milani, L., Osualdella, D., & Di Blasio, P. (2009). Quality of interpersonal relationships and problematic Internet use in adolescence. *CyberPsychology & Behavior, 12*(6), 681–684. Retrieved from Academic Search Complete database.

Newhouse, P. C. (2002). *The impact of ICT on Learning and Teaching: A literature review for the Western Australian Department of Education*. Perth: Specialist Educational Services.

Pauleen, D. (2001). Facilitating virtual team relationships via Internet and conventional communication channels. *MCB UP Ltd, 11(3)*, Retrieved from http://www.emeraldinsight.com/Insight/viewContentItem.do?contentType=Article&hdAction=lnkhtml&contentId=863712&StyleSheetView=Text

Pierce, T. (2009). Social anxiety and technology: Face-to-face communication versus technological communication among teens. *Computers in Human Behavior,* 1367–1372.

Scalet, S. D. (2007). Retrieved June 4, 2010, from http://www.csoonline.com/article/221156/A_MySpace_Time_Line

Selwyn, N. (2009). Faceworking: Exploring students' education-related use of Facebook. *Learning, Media and Technology, 34*(2), 157–174.

StatsCan. (2005). *Statistics Canada* Retrieved June 2, 2010, from http://www.statcan.gc.ca/pub/56f0004m/2005012/ch1-eng.htm

Sussman, G. (1999). Urban congregations of capital and communications: Redesigning social and spatial boundaries. *Social Text, 17*(3), 35–51.

Weber, S. (2007). *The image and identity research collective*. Retrieved June 15, 2010, from http://iirc.mcgill.ca/txp/?s=Methodology&c=Art-based%20research

Wellman, B., Quan-Haase, A., Boase, J., Chen, W., Hampton, K., Isla de Diaz, I., &Miyata, K. (2003). The social affordances of the Internet for networked individualism. *The Journal of Computer Mediated Communication, 8*(3).

YorkU.ca. (2010).*York University:, social networking theory*. Retrieved July 4, 2010, from http://www.fsc.yorku.ca/york/istheory/wiki/index.php/Social_network_theory

FURTHER READING

Abes, E., & Jones, S. (2004). Enduring influences of service-learning on college students' identity development. *Journal of College Student Development, 45*(2) 149–166.

Johnson, D., & Johnson, R. (1998). *Cooperative learning and social interdependence theory. Social Psychological Applications To Social Issues*. Retrieved July 26, 2010, from http://www.co-operation.org/pages/SIT.html

Kraut, R. (2001). Internet paradox revisited, 1–25. *Journal of Social Issues, 20,* 234–245.

Lee, Y.-C., & Chung Sun, Y (2009). Using instant messaging to enhance the interpersonal relationships of Taiwanese adolescents. *Adolescence,* 199–208.

Merkle, E., & Richardson, R. (2000). Digital dating and virtual relating: Conceptualizing computer mediated romantic relationships. *Family Relations, 49,* 187–192. JSTOR. Retrieved June 22, 2010, from http://www.jstor.org.proxy.bib.uottawa.ca/stable/585815?cookieSet=1

Pauleen, D., & Yoong, P. (2001). Relationship building and the use of ICT in boundary-crossing virtual teams: A facilitators perspective. [Electronic version]. *Journal of Information Technology, 16,* 205–220.

Raija-Leena Punamäki, M. W.-H. (2009). The associations between information and communication technology (ICT) and peer and parent relations in early adolescence. *International Journal of Behavioral Development,* 556–565.

Wilson, B. (2006). Ethnography, the Internet, and youth Culture: Strategies for examining social resistance and "Online-Offline" relationships. [Electronic version]. *Canadian Journal of Education, 29,* 307–328.

10

Bagnoli, A. (2009). Beyond the standard interview: The use of graphic elicitation and arts-based methods. *Qualitative Research, 9,* 547–

570. Retrieved July 29, 2010. from http://qrj.sagepub.com/content/9/5/547.abstract

Butler-Kisber, L. (2008). Collage as Inquiry. In J. Knowles & A. Cole (Eds.), *Handbook of the arts in qualitative research: Perspectives, methodologies, examples, and issues* (pp. 265–276). Thousand Oaks, CA: Sage.

Butler-Kisber, L. (2010). *Qualitative inquiry, thematic, narrative & arts-informed perspectives*. California, CA: Sage.

Boyd, D. (2008). Why youth love social network sites: The role of networked publics in teenage social life. In D. Buckinham, (Ed.), *Youth, Identity, and Digital Media* (119–142). The John D. and Catherine T. MacArthur foundation Series on Digital Media and Learning. Cambridge, MA: The MiT Press. Retrieved July 29, 2010, from http://papers.ssrn.com/sol3/papers.cfm?abstract_id=1518924

Boyd, D. M., & Ellison, N. B. (2008). Social network sites: Definition, history, and scholarship, *Journal of Computer-Mediated Communication, 13*(1), 210–230. Retrieved July 29, 2010, from http://www3.interscience.wiley.com/cgi-bin/fulltext/117979376/PDFSTART

Creswell, J. (2007). *Qualitative inquiry & research design: Choosing among five approaches*. California, CA: Sage.

Carroll, J., Howard, S., Vetere, F., Peck, J., & Murphy, J. (2002). *Just what do the youth of today want? Technology appropriation by young people*. Proceedings of the 35th Hawaii International Conference on System Sciences. Retrieved July 28, 2010, from http://www.computer.org/portal/web/csdl/doi/10.1109/HICSS.2002.994089

Ellison, N. B., Steinfield, C., & Lampe, C., (2007). The benefits of Facebook "Friends:" Social capital and college students' use of online social network sites. *Journal of Computer-Mediated Communicaiton, 12*, 1143–1168. Retrieved July 29, 2010, from http://www3.interscience.wiley.com/cgi-bin/fulltext/117979349/PDFSTART

Galman, S. (2009). The truthful messenger: Visual methods and representation in qualitative research and education. *Qualitative Research, 9*(2), 197–217. Retrieved July 29, 2010, from http://qrj.sagepub.com/content/9/2/197.full.pdf+html

Gauntlett, D. (2004). *Using New Creative Visual Research Methods to Understand the Place of Popular Media in People's Lives*. Paper presented for the International Association for Media and Communicaiton Research, Porto Alegre, Brazil. Retrieved July 29, 2010, from http://www.artlab.org.uk/iamcr2004.htm

Gonzales, A., & Hancock, J. (2008). Identity shift in computer-mediated environments. *Media Psychology, 11*(2), 167–185. Retrieved July 28, 2010, from http://pdfserve.informaworld.com/529904_770885140_794359855.pdf

Groenewald, T. (2004). A phenomenological research design illustrated. *International Journal of Qualitative Methods, 3*(1), 1–26. Retrieved July 28, 2010, from http://www.ualberta.ca/~iiqm/backissues/3_1/html/groenewald.html

Joinson, A. N. (2001). Self-disclosure in computer-mediated communication: The role of self-awareness and visual anonymity. *European Journal of Social Psychology, 31*, 177–192. Retrieved July 29, 2010, from http://www3.interscience.wiley.com/cgi-bin/fulltext/78503088/PDFSTART

Merchant, G. (2006). Identity, social networks and online communication. *E-Learning, 3*(2). Retrieved July 29, 2010, from http://www.ww-words.co.uk/pdf/validate.asp?j=elea&vol=3&issue=2&year=2006&article=9_Merchant_ELEA_3_2_web

Office of the Privacy Commissioner of Canada. (March 20, 2009). *Focus testing privacy issues and potential risks of social networking sites* (Research Report). Retrieved July 20, 2010, from http://www.priv.gc.ca/information/survey/2009/decima_2009_02_e.cfm

Siegesmund, R. (2008). *"Visual Research" The Sage Encyclopedia of Qualitative Research Methods*. Retrieved July 17, 2010, from http://www.sage-ereference.com/research/Article_n491.html?searchQuery=y%3D12%26searchPage%3D0%26quickSearch%3Dmethodologies%2Bto%2Banalyze%2Bvisual%2Bdata%2B%26x%3D32

Stritzke, W. G. K., Nguyen, A., & Durkin, K. (2004). Shyness and computer-mediated communication: A self-presentational theory perspective. *Media Psychology, 6*, 1–22. Retrieved July 8, 2010, from http://pdfserve.informaworld.com/205831_770885140_785350902.pdf

Stryker, S., & Burke, P. J. (2000). The past, present, and future of an identity theory. *Social Psychology Quarterly, 63*(4), 284–297. Retrieved July 29, 2010, from http://www.jstor.org/stabl e/2695840?&Search=yes&term=%22The+Pa st%2C+Present%2C+and+Future+of+an+Ide ntity+theory%22&list=hide&searchUri=%2F action%2FdoBasicSearch%3FQuery%3D%25 22The%2BPast%252C%2BPresent%252C%2 Band%2BFuture%2Bof%2Ban%2BIdentity% 2Btheory%2522%26wc%3Don%26dc%3DA ll%2BDisciplines&item=1&ttl=39&returnAr ticleService=showArticle

Suler, J. R. (2002, October). Identity management in cyberspace, *Journal of Applied Psychoanaytic Studies, 4*(4), 455–460. Retrieved July 28, 2010, from http://www.springerlink.com/ content/l322287n68344640/fulltext.pdf

Turkle, S. (1999). Cyberspace and identity. *Contemporary Sociology, 28*, 643–648.

Weber, S. (2008). Visual images in research. In. J. Knowles & A. Cole (Eds.), *Handbook of the arts in qualitative research: Perspectives, methodologies, examples, and issues* (pp. 41–53). Thousand Oaks, CA: Sage.

Zhao, S., Grasmuck, S., & Martin, J. (2008, September). Identity construction on Facebook: Digital empowerment in anchored relationships. *Computers in Human Behavior, 24*(5), 1816–1836. Retrieved July 28, 2010, from http://www.sciencedirect.com/ science?_ob=MImg&_imagekey=B6VDC-4S2VRJ0-1-5&_cdi=5979&_user=1067359&_ pii=S0747563208000204&_orig=search&_ coverDate=09%2F30%2F2008&_ sk=999759994&view=c&wchp=dGLzVlb-zSkz k&md5=9eae4acd5516e4662c2206d4cd332a90 &ie=/sdarticle.pdf

FURTHER READING

Kennedy, H. (2006). Beyond anonymity, or future directions for internet identity research. *New Media Society, 8*, 859–879.

Mesch, G., & Talmud, I. (2006). The quality of online and offline relationships: the role of multiplexity and duration of social relationships. *The Information Society, 22*, 137–148.

Ramirez, A., & Zhang, S. (2007). When online meets offline: the effect of modality switching in relational communication. *Communication Monographs, 74*(3), 287–310.

Ribeiro, J. C. (2009). The increase of the experience of self through the practices of multiple virtual identities. *Psychology Journal, 7*(3), 291–302.

Suler, J. (2004). The online disinhibition effect. *Cyberpsychology & Behaviour, 7*, 321–326.

Walther, J. B. (1996). Computer-mediated communication: Impersonal, interpersonal, and hyperpersonal interaction. *Communication Research, 23*, 3–43.

Wang, Z., Walther J. B., & Hancock, J. T. (2009). Social identification and interpersonal communication in computer-mediated communication: What you do versus who you are in virtual groups. *Human Communication Research, 35*, 59–85.

11

Bargh, J. A., McKenna, A. K., & Fitzsimons, M. G. (2002). Can you see the real me? Activation and expression of the "true self" on the Internet. *Journal of Social Issues*, 33–48.

Creswell, J. (2007). *Qualitative inquiry and research design: Choosing among five approaches*. California: Sage.

Creswell, J. (2009). *Research design: Qualitative, quantitative, and mixed methods approaches*. Los Angeles, CA: Sage.

Froomkin, M. (1996). Flood control on the information ocean: Living with anonymity, digital cash and distributed databases. *Journal of Law and Commerce*, 395–515.

Gies, L. (2008). How material are cyberbodies? Broadband Internet and embodied subjectivity. *Crime Media Culture*, 311–330.

Gonzales, A. L., & Hancock, J. T. (2008). Identity shift in computer-mediated environments. *Media Psychology*, 167–185.

Kiesler, S., Siegel, J., & McGuire, T. (1984). Social psychological aspects of computer-mediated communication. *American Psychologist*, 1123–1134.

Moral-Toranzo, F., Canto-Ortiz, J. & Gomez-Jacinto, L. (2007). Anonymity effects in computer-mediated communication in the case of minority influence. *Computers in Human Behavior,* 1660–1674.

Morio, H., & Buchholz, C. (2009). How anonymous are you online? Examining online social behaviors from a cross-cultural perspective. *AI & Society,* 297–307.

Moustakas, C. (1994). *Phenomenological research methods.* Thousand Oaks, CA: Sage.

Nissenbaum, H. (1999). The meaning of anonymity in an information age. *The Information Society,* 141–144.

Parks, M., & Floyd, K. (1996). Making friends in cyberspace. *Journal of Communication,* 97.

Pauwels, L. (2005). Websites as visual and multinodal cultural expressions: Opportunities and issues of online hybrid media research. *Media, Culture and Society,* 604–611.

Pfitzmann, A., & Hansen, M. (2008, February). *Anonymity, unlinkability, unobservability and pseudonymity, and identity management—a consolidated proposal for terminology v0.31.* Retrieved from http://dud.inf.tu-dresden.de/literatur

Rosenberg, M. (1986). *Conceiving the self.* Malabar, FL: Robert E. Krieger.

Turkle, S. (1996). *Life on the screen: Identity in the age of the Internet.* London: Weidenfeld & Nicolson.

Vieta, M. A. (2004). Interactions through the screen: The interactional self as a theory for internet-mediated communication. Simon Fraser University.

Walther, J. B. (1992). Interpersonal effects in computer-mediated interaction: A relational perspective. *Communication Research, 19,* 52–90.

Woo, J. (2006). The right not to be identified: Privacy and anonymity in the interactive media environment. *New Media & Society,* 949–967.

Zhao, S., Grasmuck, S., & Martin, J. (2008). Identity construction on Facebook: Digital empowerment in anchored relationships. *Computers in Human Behavior,* 1816–1836.

FURTHER READING

Festinger, L., Pepitone, A., & Newcomb, T. (1952). Some consequences of deindividuation in a group. *Journal of Abnormal and Social Psychology,* 382–389.

Gergen, K. J., Gergen, M. M., & Barton, N. W. (1973). Deviance in the dark. *Psychology Today,* 129–130.

Housley, S. (2006). The dangers of an anonymous Internet. Retrieved from Ezine @rticles: http://ezinearticles.com/?The-Dangers-of-an-AnonymousInternet&id=267988

Kennedy, H. (2006). Beyond anonymity, or future directions for internet identity research. *New Media Society,* 859–876.

Kioskea. (2009, January, 14). *Social network use by adult Americans on the rise: Survey.* Retrieved from http://en.kioskea.net/news/11805-social-network-use-by-adult-americans-on-the-rise-survey

Kokswijk, J. V. (2008). Granting personality to a virtual identity. *International Journal of Human and Social Sciences,* 207–215.

Lea, M. (1998). *Effects of visual anonymity in computer-mediated group interactions.* Retrieved from http://virtualsociety.sbs.ox.ac.uk/results/lea.pdf

Lea, M., Spears, R., & de Groot, D. (2001). Knowing me, knowing you: Anonymity effects on social identity processes within groups. *Personality and Social Psychology Bulletin,* 526–537.

MacLeod, K. (1999). *The online identity: How MUDs shape fantasy into reality.* Retrieved from Building the Virtual City: Suggestions for Shaping a Viable Cybersociety: http://socserv.mcmaster.ca/soc/courses/stpp4C03/ClassEssay/muds.htm

Nosko, A., Wood, E., & Molema, S. (2010). All about me: Disclosure in online social networking profiles: The case of FACEBOOK. *Computers in Human Behavior,* 406–418.

Postmes, T., & Spears, R. (1998). Deindividuation and antinormative behavior: A meta-analysis. *Psychological Bulletin,* 238–259.

Reicher, S., Spears, R., & Postmes, T. (1995). A social identity model of Deindividuation phenomena. *European Review of Social Psychology,* 161–199.

Schmitz, J. (1997). Structural relations, electronic media, and social change: The Public electronic network and the homeless. In S. Jones (Ed.), *Virtual culture: Identity and communication in cyberspace* (pp. 80–101). London: Sage.

Schwartz, P. (1999). Privacy and democracy in cyberspace. *Vanderbilt Law Review*, 1609–1705.

Shin, H. K., & Kim, K. K. (2010). Examining identity and organizational citizenship behaviour in computer-mediated communication. *Journal of Information Science*, 114–126.

Skok, G. (2000). Establishing a legitimate expectation of privacy in clickstream data. *Michigan Telecommunications & Technology Law Review*, 61–85.

Stern, S. R. (2004). Expressions of identity online: Prominent features and gender differences in adolescents' World Wide Web home pages. *Journal of Broadcasting & Electronic Media*, 218–243.

Tanis, M., & Postmes, T. (2007). Two faces of anonymity: Paradoxical effects of cues to identity in CMC. *Computers in Human Behavior*, 955–970.

Tobeme. (2007). *Teenagers—Finding their identity*. Retrieved October 29, 2007, from http://tobeme.wordpress.com/2007/10/29/teenagers-finding-their-identity/

Walther, J. B. (1996). Computer-mediated communication: Impersonal. *Interpersonal and Hyperpersonal*, 3–43.

Walther, J. B., & Burgoon, K. J. (1992). Relational communication in computer-mediated interaction. *Human Communication Research*, 50–88.

12

Bailey, F. (1969). *Stratagems and spoils: A social anthropology of politics*. New York: Blackwell.

Dyson, E., Gilder, G., Keyworth, G., & Toffler, A. (1994). *Cyberspace and the American dream: A magna carta for the knowledge age*. Future Insight 1.2. The Progress & Freedom Foundation.

Luppicini, R. (2008). Introducing technoethics. In R. Luppicini & R. Adell (Eds.), *Handbook of research on technoethics* (pp. 1–18). Hershey, PA: Idea Group Publishing.

Luppicini, R. (2010). *Technoethics and the evolving knowledge society*. Hershey, PA: Idea Group Publishing.

Pinch, T. J., & Bijker, W. E. (1987). *The social construction of technological systems: New directions in the sociology and history of technology*. Cambridge, MA: MIT Press.

University of Ottawa Vision 2020. (2010). *Strategic plan vision 2020*. Retrieved November 1, 2010, from, http://strategicplanning.uottawa.ca/vision2020/committee.html

FURTHER READING

Bell, D. (1973). *The coming of post-industrial society: A venture in social forecasting*. New York: Basic Books.

Bunge, M. (1977). Towards a technoethics. *Monist, 60*(1), 96–107.

Bunge, M. (1979). A systems concept of society: Beyond individualism and holism. *Theory and Decision, 10*(1), 13–30.

Mohr, H. (1999). Technology assessment in theory and practice. *Techné: Journal of the Society for Philosophy and Technology, 4*(4), 1–4.

Schumacher, E. (1973). *Small is beautiful*. New York: Harper & Row.

Seemann, K. (2003) Basic principles in holistic technology education. *Journal of Technology Education, 14*(2), 12–24.

Shrader-Frechette, K. (1997). Technology and ethical issues. In K. Shrader-Frechette & L. Westra (Eds.), *Technology and values* (pp. 25–32). Lanham, MD: Rowman & Littlefield.

Simon, J. (Ed.). (1995). *The state of humanity*. Cambridge, MA: Blackwell.

Strijbos, S., & Basden, A. (Eds.). (2006). *In search for an integrative vision for technology*. Dordrecht: Springer.

Toulmin, S. (1958). *The uses of argument*. Boston, MA: Cambridge University Press.

Verbeek, P. (2005). *What things do: Philosophical reflections on technology, agency, and design*. In R. P. Crease (Trans.). University Park, PA: Pennsylvania State University Press.

Verbeek, P. (2006). Materializing morality: Design ethics and technological mediation. Materializing Morality. *Science, Technology, & Human Values, 31*(3), 361–380.

Wajcman, J. (1991) *Feminism confronts technology*. University Park, PA: Pennsylvania State University Press.

Index